SpringerBriefs in Computer Science

Series Editors
Stan Zdonik
Peng Ning
Shashi Shekhar
Jonathan Katz
Xindong Wu
Lakhmi C. Jain
David Padua
Xuemin Shen
Borko Furht
VS Subrahmanian

T0211648

For further volumes:
http://www.springer.com/series/10028

Akrivi Vlachou • Christos Doulkeridis
Kjetil Nørvåg • Yannis Kotidis

Peer-to-Peer Query Processing over Multidimensional Data

Akrivi Vlachou
Norwegian University of Science
    and Technology
Trondheim, Norway

Christos Doulkeridis
Norwegian University of Science
    and Technology
Trondheim, Norway

Kjetil Nørvåg
Norwegian University of Science
    and Technology
Trondheim, Norway

Yannis Kotidis
Athens University of Economics
    and Business
Athens, Greece

ISSN 2191-5768   ISSN 2191-5776 (electronic)
ISBN 978-1-4614-2109-2   ISBN 978-1-4614-2110-8 (eBook)
DOI 10.1007/978-1-4614-2110-8
Springer New York Heidelberg Dordrecht London

Library of Congress Control Number: 2012936124

Printed on acid-free paper

Springer is part of Springer Science+Business Media (www.springer.com)

Preface

Nowadays, applications that require a high degree of distribution and loosely-coupled connectivity are ubiquitous in various domains, including scientific databases, bioinformatics, and multimedia retrieval. In all these applications, data is typically voluminous and multidimensional, and support for advanced query operators is required for effective querying and efficient processing. Moreover, the highly distributed setting calls for a system architecture with salient properties, including scalability, fault-tolerance, autonomy, and dynamic participation.

Peer-to-peer (P2P) systems emerge as a powerful model for searching vast amounts of data distributed over independent sources. Each peer stores autonomously its own data and the objective is to support efficient and effective techniques for query processing and advanced data analysis. In this context, a key requirement of query processing is retrieval of the exact and complete result set. To achieve this goal, a scalable framework is presented that relies on data summaries that are distributed and maintained as multidimensional routing indices. Different types of data summaries enable efficient processing of a variety of advanced query operators. This book focuses on queries for similarity search, skyline queries, and top-k queries, and identifies appropriate data summaries, proposes effective indexing methods at local level as well as routing indices at network level, and efficient processing algorithms for each query type.

Akrivi Vlachou
Christos Doulkeridis
Kjetil Nørvåg
Yannis Kotidis

Contents

About the Authors

Akrivi Vlachou is a post-doctoral researcher at the Norwegian University of Science and Technology (NTNU) in collaboration with Athena Research and Innovation Center, Athens, Greece. She received her Ph.D. in 2008 from the Athens University of Economics and Business (AUEB), her MSc degree and her B.Sc. degree from the Department of Computer Science and Telecommunications of University of Athens in 2003 and 2001 respectively. In her dissertation, she studied methods for efficient query processing for highly distributed data. She has received fellowships for post-doctoral studies from European Research Consortium for Informatics and Mathematics (ERCIM) and from the Greek State Scholarship Foundation. Her research interests include query processing and data management in distributed systems, algorithms and query operators for large-scale data analysis and spatial-keyword search over web-accessible data.

Christos Doulkeridis is a post-doctoral researcher at the Norwegian University of Science and Technology. He has previously been awarded with an ERCIM *Alain Bensoussan* fellowship for post-doctoral studies. He holds a Ph.D. and a MSc in Information Systems from the Department of Informatics of Athens University of Economics and Business and a B.Sc. degree in Electrical Engineering and Computer Science from the National Technical University of Athens. His research interests include P2P and distributed data management, cloud computing, distributed knowledge discovery, mobile and context-aware computing.

Kjetil Nørvåg is a Professor in the Department of Computer and Information Science at the Norwegian University of Science and Technology. He received a Dr.Ing. degree in computer science from the Norwegian University of Science and Technology in 2000. He has been a visiting researcher at INRIA in Paris, Athens University of Economics and Business, and Aalborg University. His major research interests include distributed database systems, information retrieval, and text mining. He has published more than 120 papers in international refereed conferences and peer reviewed journals.

Yannis Kotidis is an Assistant Professor in the Department of Informatics of Athens University of Economics and Business. He holds a B.Sc. degree in Electrical Engineering and Computer Science from the National Technical University of

Athens, an MSc and a Ph.D. in Computer Science from the University of Maryland (USA). Between 2000 and 2006 he was a Senior Technical Specialist at the Database Research Department of AT&T Labs-Research in Florham Park, New Jersey. His main research areas include large scale data management systems, data warehousing and sensor networks.

Chapter 1
Introduction

Abstract During the last decades, the vast number of independent data sources and the high rate of data generation make central assembly of data at one location infeasible. As a consequence, data management and storage become increasingly distributed. Although the client-server architecture as communication model is still popular for application development, traditional client-server solutions are prone to bottleneck risks and therefore do not scale as the number of participating sources increases. Peer-to-peer (P2P) systems comprise a scalable collaborative architecture of autonomous participants (called *peers*), where each peer serves both as a client and as a server. As data storage becomes inherently distributed, an emerging challenge is to support efficient query processing over data stored at disparate locations, which allows users to discover and retrieve relevant data. Even more important is to incorporate and support flexible query operators, such as similarity search, skyline and top-k queries, that help avoiding huge and overwhelming result sets. Moreover, as data in modern applications is typically multidimensional, effective distributed query processing methods and techniques that apply to multidimensional data are sought. Query processing in P2P networks poses inherent challenges and demands non-traditional techniques due to the distributed nature of the environment and lack of global knowledge. This chapter provides an introduction to the main topics and basic concepts related to P2P query processing over multidimensional data.

1.1 Peer-to-Peer Systems

Peer-to-peer (P2P) systems emerge as a powerful model for searching huge amounts of data distributed over independent sources in a completely distributed and self-organizing way (see [4, 107] for comprehensive surveys). Peers are processing units with storage capabilities that act both as a client and as a server. P2P systems are classified in *unstructured* and *structured* P2P systems, based on the way the content is located in the network, with respect to the network topology; unstructured P2P systems are loosely-connected and do not impose any constraints on the way peers are connected, whereas in structured P2P systems peers are organized in a more rigid topology. In addition, P2P systems are distinguished in purely decentralized, hybrid centralized and hybrid decentralized, based on the degree of centralization; in pure P2P systems all peers perform the same tasks without any central coordination, whereas hybrid systems include some form of centralization, either by one peer (hybrid centralized) or a set of peers (hybrid decentralized).

Attractive features of P2P networks include scalability to large numbers of peers, resilience to failures, loose coupling, and peer autonomy. However, the high degree of distribution comes at a cost; efficient search and query processing mechanisms need to be established to decrease the cost of data retrieval. Even though different P2P architectures have been employed for establishing query processing mechanisms, in this book, the focus is on an unstructured, hybrid decentralized P2P architecture.

1.2 Multidimensional Data Management

In several real-life applications, including scientific data management, multimedia retrieval, and digital libraries, data is typically multidimensional. Querying multidimensional data raises significant challenges relevant to efficiency, even in centralized domains. Specialized multidimensional access methods are required to support efficient query processing, and there exists a rich literature about managing multidimensional data [116].

The aim of multidimensional data management is to process efficiently a large number of data objects, where each data object is characterized by a set of features that define a multidimensional data space. In this multidimensional data space, each data object is represented by a point, where the coordinates of the data point correspond to the values of the features. In applications that involve multidimensional data, it is very common that instead of exact match queries, the user is interested in posing more flexible query operators that allow the discovery of the most interesting data objects in a large data collection. Examples of such queries are similarity search, top-k and skyline queries. These advanced query types are more challenging than exact match queries and effective access methods are required for efficient query processing.

1.3 Peer-to-Peer Query Processing

Assume a P2P system that consists of a set of autonomous peers, each maintaining a local database that conforms to a common schema. If one could collect all data at one central location, then this centralized database would have the same schema as each local database of an individual peer, only the data objects would be the union of all data objects located at all peers. This is the case of *horizontal distribution* of data to peers.

Query processing over the P2P system retrieves a set of data objects, also mentioned as query result set, that satisfy the query constraints. The aim of distributed query processing is to retrieve the same query result as if the query has been posed to the centralized database of all data objects. P2P query processing refers to the algorithmic steps that are employed for retrieval of the query result set. The main tasks of P2P query processing that influences the performance of searching are:

- *Local query processing* at each peer that undertake the task of query processing over the peer's local database only.
- *Query routing* that aim at forwarding the query to those peers that maintain local data that belong to the query result set.

In the former case, traditional query processing algorithms are usually applicable with minor modifications. In the latter case, P2P query processing algorithms need an efficient routing mechanism that deliberately forwards queries in such a way that target peers (those with query results) are reached in a cost-effective manner. At the same time, the routing mechanism should not forward the query to peers that cannot contribute to the query result set. The routing information used that is built and maintained at each individual peer needs to aggregate information about what data is accessible when a certain direction is followed in the network. This corresponds essentially distributed routing mechanism, which is not straightforward to build and demands non-traditional techniques due to the distribution of content.

1.4 Organization

The focus of this book is efficient query processing of multidimensional data in P2P systems, where the data objects are horizontally distributed over the peers. The rest of the book is organized as follows: Chapter 2 provides useful background information on basic P2P concepts. Then, Chapter 3 presents a clear overview of the system model adopted, as well as an overall description of the framework for P2P multidimensional query processing and routing. Chapter 4 provides definitions of advanced query operators that are the focus of the book. Thereafter, Chapter 5 presents similarity search in P2P systems, focusing on range queries. Chapter 6 describes P2P skyline query processing, with special focus on subspace skyline queries. Chapter 7 provides details on supporting top-k queries in P2P systems. Finally, Chapter 8 summarizes the most important findings and outlines open research problems.

Chapter 2
Peer-to-Peer Systems

Abstract In this chapter, a basic overview is given of P2P systems, architectures, and search strategies in P2P systems. More specific concepts that are outlined include the differences of structured and unstructured P2P systems, categories of P2P systems based on the centralization degree, basic search mechanisms for unstructured P2P systems, as well as details on P2P data management systems.

2.1 General Characteristics

P2P systems refer to distributed computer architectures that are designed for sharing resources, by direct exchange, rather than requiring a central coordinating server. In P2P systems, the interconnected computers, called *peers*, are organized in a network in a distributed and self-organizing way and share resources, while respecting the autonomy of peers. The main characteristic of a P2P system is the ability to adapt to peer failures (fault-tolerance) and accommodate a large number of participating peers (scalability), while keeping the performance of the network at an acceptable level and maintaining the peer connectivity.

Generally, all peers are equivalent in terms of tasks and functionality they perform, and act both as clients and servers at the same time. Peers participate in the system in a collaborative manner, by performing tasks such as connecting to other peers, routing messages to other peers and searching for content. With respect to the peer autonomy, each peer maintains and controls its own content or resources.

Even though different kinds of resources may be shared over a P2P network, such as CPU or storage, P2P systems emerge as a powerful model for organizing and searching of huge amounts of data distributed over independent sources. Applications such as file-sharing, distributed database and digital libraries gain from such an architecture. Therefore, in this book, we focus on P2P systems that share content, one of the most prominent applications areas of P2P technology.

Each peer has a collection of files or data to share. Two peers that maintain an open connection between them are called *neighbors*. The number of neighbors of a node defines its *outdegree* or *degree*. Any peer can pose a query, in order to retrieve interesting content; this peer is called *querying peer*. When a peer receives a query, first the query is evaluated against its local data collection and thereafter, if necessary other peers are contacted through its neighbors. Query messages are forwarded only between open connections, i.e., neighboring peers. If a message has to be exchanged between two non-neighboring peers, more than one message is required. The cost of this message is considered that it is proportional to the length of the path, also called number of *hops*. Usually query results are forwarded back to the querying peer through the reverse path of the query message. Alternatively, some approaches allow the establishment of a temporary connection, in order to transfer the query results directly to the querying peer.

In the rest of this chapter, we will describe in more detail different architectures for P2P systems, how to search for data in such systems, and finally we give an overview of some P2P-based data management systems.

2.2 P2P Architectures

P2P systems can be classified into two categories, based on the way the content is located in the network, with respect to the network topology: *unstructured* and *structured*. In unstructured P2P, each peer maintains a limited number of connections (also called links) to other neighboring peers in the network. Searching in an unstructured P2P environment usually leads to either flooding queries in the network using a time-to-live (TTL) or query forwarding based on constructed routing indices that give a direction for the search. Examples of such unstructured P2P networks include Gnutella and Freenet [39]. In structured P2P systems, a hash function is used in order to couple keys with objects. Then a distributed hash table (DHT) is used to route key-based queries efficiently to peers that hold the relevant objects. In this way, object access is guaranteed within a bounded number of hops. Examples of popular structured P2P systems are Chord [121], CAN [106], Pastry [112], Tapestry [146].

2.2.1 Structured P2P Systems

Structured P2P systems assume a relation between the peer content and the network topology. The topology of the network is strongly controlled and data is stored or indexed by specific peers. Therefore, there is a mapping between the data objects and the location, i.e., the peer that stores or indexes it, so that queries can be routed easily to peers that store relevant data objects. Usually the mapping is realized as a distributed hash table (DHT). We shortly describe two representative structured P2P networks, namely CHORD [121] and CAN [106].

CHORD [121] is a P2P protocol that uses a distributed hash table for efficiently locating the peer that stores the data item corresponding to a given one-dimensional key value. CHORD uses *consistent hashing* to map uniformly the domain of search keys into the Chord domain of keys. More precisely, each peer is identified by an *m*-bit key value generated by the hash function applied on the peer's IP. Then, the peers are ordered in an *identifier circle* based on their key value. Each object with a key value k is assigned to the first peer that has a key value equal to or larger than k. In order to maintain efficient communication between peers, every peer stores the addresses of its predecessor and successor on the identifier circle. In addition, each peer maintains a routing table called the *finger table* with addresses of up to m other nodes. With high probability, the peer responsible for a given key value is located via $O(log(N))$ number of messages to other peers.

CAN [106] divides dynamically the d-dimensional data space into partitions, so that each peer is assigned to a data space partition. Each peer connects to the peers that store neighboring data space partitions, according to some dimension. Each time a peer joins the network, it randomly chooses a point in the d-dimensional space and contacts the peer responsible for the partition in which that data points belongs. Then this peer, splits the data space in two parts, based on some dimension and then assigns the one part to the new peer. Finally, the neighboring peers are contacted and updated. A point query based on the CAN architecture is performed in $O(N^{1/d})$ messages.

2.2.2 Unstructured P2P Systems

In contrast to structured P2P networks, in unstructured P2P networks peers connect to other peers in a random way and there is no relation of the placement of the data objects with the network topology. Therefore, peers have no or limited information about data objects stored by other peers, thus searching in unstructured P2P networks may lead to query all neighbors for data items that match the query (see Section 2.3).

Unstructured P2P networks are easy to implement and impose low maintenance cost, however query routing is not scalable. As the number of participant peers increases, the number of exchanged messages increases too. On the other hand, structured P2P networks maintain information about the data objects available through their neighbors. Therefore, query processing algorithms with cost bounds have been proposed. Nevertheless, the maintenance cost is high, especially for large P2P networks and high rates of peer joins or departures (also known as *churn*).

2.2.3 P2P Network Centralization Degree

P2P overlay networks were initially defined to be totally decentralized, however for practical reasons, systems with different degrees of centralization have been developed. Therefore, except of purely decentralized architectures, also hybrid systems were proposed.

In a purely distributed P2P system, all nodes in the network perform exactly the same tasks, acting both as clients and servers, and there is no central coordination of their activities. The major shortcomings of purely distributed systems is scalability issues and the poor performance during query processing.

In the hybrid centralized indexing systems, there is a central server facilitating the interaction between peers and a centralized index is build at this specialized peer. Napster, for example, uses a centralized index, that is built in cooperation with all the participating peers. The centralized index keeps information about the data stored at each peer, together with the peer identifier. Therefore, centralized indices are efficient during query processing; a single message is required to determine which peer stores relevant information. Notice that the actual sharing of information between peers is established by communication between the peers, without interaction with the central server. Despite the efficiency during query processing, centralized indices have a major drawback, namely they constitute a "single point of failure". Moreover, the centralized index may become a bottleneck for the system, especially in the case of a large P2P network.

Hybrid decentralized indexing systems, namely super-peer infrastructures [141], harness the merits of both centralized and purely distributed architectures. Super-peer networks tackle the scaling and "single-point-of-failure" problems of centralized approaches, while exploiting the advantages of the completely distributed approach, where each peer builds and maintains an index over its own files. These systems are similar to purely decentralized systems, but some of the peers have a more important role, and are responsible to maintain the information available at their associated peers and facilitate the interaction between peers. If only the super-peers are considered, they act as a purely decentralized P2P system.

2.3 Search in Unstructured P2P Systems

Different approaches have been proposed for unstructured P2P systems, in order to find interesting content for the users.

2.3.1 Flooding-Based Search

The most naive approach that does not use any index is *flooding* the network, until interesting content is retrieved. Flooding is performed by each peer forwarding the query to all neighbors, except from the one that the query was received. It is obvious, that flooding-based search does not constitute a truly scalable solution, as the query-induced traffic practically saturates the P2P system. Therefore, several variations have been proposed, in order to reduce the cost of flooding.

An alternative to pure flooding is to attach a *time-to-live* (TTL) that specifies after how many hops the query should be stopped, which leads to the problem of a limited horizon. Any peer receiving a query, first decrements the TTL and only if the TTL is greater than zero it forwards the query to its neighbors. Flooding was used by Gnutella. Although this approach is simple and there is no overhead for maintaining routing information, it has the disadvantage that it may lead to an enormous cost during query processing, especially in the case of large P2P networks. In addition, flooding is sensitive to the connectivity degree of each peer, since a small connectivity degree leads to long routing paths, while in a dense network topology each peer receives the query from multiple paths, resulting in a large number of exchanged messages and waste of bandwidth.

Another variant is to perform a *random walk* on the network [60]. Using this strategy, the query is forwarded to one or some random neighbors instead of forwarding to all neighboring peers.

2.3.2 Routing Indices

Routing indices [42] were proposed to alleviate the shortcomings of the aforementioned search methods. Routing indices are implemented as multiple small indices that are stored distributed at every peer, that differs to a costly and vulnerable centralized index. Routing indices store the direction that should be followed, in order to find the requested data, rather than the corresponding data itself. In more detail, each peer maintains an index that describes the information that is available through each neighboring peer. For example, in the case of documents, a centralized index would store the terms with the peers' IPs that store documents containing the terms, while a routing index would maintain only information about how many documents containing each term we could find by following a path starting at each neighboring peer. Therefore, each peer is able to forward a query to the neighbors that are more likely to have results, instead of choosing some neighbors randomly or flooding the network by forwarding the query to all neighbors. The routing indices are usually constructed assuming an acyclic network topology, while different approaches are proposed for handling cycles.

2.3.3 Semantic Overlay Networks

In application areas were data is non-uniformly distributed over the peers, and it is possible to discover peers with similar content, then it is possible to form thematically focused peer groups and use this for query routing. This technique, called *semantic overlay networks* [53], makes it possible to route queries to only those peer groups that are relevant to the query, thus significantly reducing the number of peers that has to be contacted during a query.

2.4 Super-Peer Systems

The choice of the underlying network architecture plays an important role in the performance of a P2P system. However, no single architecture is suitable for all types of possible P2P applications. Super-peer infrastructures [141] harness the merits of both centralized and distributed architectures, as they tackle both the scaling limitations and the "single-point-of-failure" problem of centralized approaches, while exploiting the advantages of the completely distributed approach, where each peer builds an autonomous index over its own files. Nowadays, super-peer architectures have been established in most of the existing P2P file-sharing applications (eMule, KaZaA).

Moreover, an important performance issue in P2P networks is uneven load distribution that can lead certain points of the network to become a bottleneck, caused by the limited capabilities of some peers. Super-peer networks take advantage of the heterogeneity of peer capabilities (e.g., bandwidth, processing power), which recent studies have shown to be enormous [117], by assigning greater responsibility to those that are more capable of handling it. In addition, in several applications, co-operative computers in a P2P network have different roles by nature, similar to the case of file-sharing, where some machines are registered as dedicated servers to the system, while others are plain personal computers that mostly request information. Furthermore, in order to respect peers autonomy, any approach should not rely on arbitrary data movement, hence each peer joining the network should autonomously store its own data. Therefore, a super-peer architecture appears particularly suitable for applications that require efficient performance for advanced query operators, hence we model our distributed system as a super-peer network.

Numerous interesting applications that require efficient query processing can be deployed over such a super-peer architecture. The overall objective is for a set of cooperative computers to collectively provide enhanced processing facilities, aiming at overcoming the limitations of centralized settings, for example extremely high computational and storage load. Examples of challenging applications that can be realized over the proposed framework include distributed image retrieval, document retrieval in digital libraries, P2P information sharing, data integration from distributed sources, as well as distributed scientific databases.

2.5 P2P-Based Data Management Systems

During the last decade a number of P2P-based data management systems have been realized. Some provides just basic storage capabilities, while others also support advanced query capabilities.

OceanStore [88] is one of the storage systems without query capabilities. It provides an infrastructure for permanent storage and replication of objects, but no query system. Objects are accessed based only on their globally unique ID, and this ID has to be known in order to retrieve or update the object.

PIER [77] is a middleware query engine built on top of a DHT. PIER does not permanently store its data. Data sources publish their data in the DHT and update them regularly, and data that are not refreshed are removed. Typically, a PIER network will contain only object metadata (e.g., filenames, sizes, tags) and a reference to the original data object. Clients will query the network to get the references to the objects of interest and retrieve the objects separately.

Among the systems that provide a full DBMS, with both query processing and storage, are Hyperion [110], Orchestra [126], Piazza [66]. All these systems allow each site to have its own schema, and use schema mediation techniques to allow cross-site querying. PeerDB [99] also falls into this category of systems with heterogeneous schemas, but the approach to schema mediation is different. Instead of relying on schema mediators, information retrieval techniques are used to find matching relations.

A slightly different approach is DASCOSA-DB [67], which does not use schema mediation. While the systems mentioned above are meant to connect existing databases and provide a common query interface, DASCOSA-DB is a distributed database system with a high degree of site autonomy, but which still behaves as one system, not many different systems with a common interface.

Other systems based on a common schema include APPA [1]. APPA provides a multilayered solution on top of a structured or super-peer P2P network, where the bottom layer is a simple key/value-store and the top level provides advanced services such as schema management, replication and query processing.

AmbientDB [20] is a system designed to provide full relational database functionality for stand-alone operation in autonomous devices that may be mobile and disconnected for long periods of time, while enabling them to cooperate in an ad hoc way with (many) other AmbientDB devices. A DHT is used both as a means for connection peers in a resilient way as well as supporting indexing of data. AmbientDB is a system for mobile devices, which have low computational power and may frequently be disconnected from the network, compared to for example DASCOSA-DB that is designed for sites that have the computational power necessary to do query processing and more stable network connections.

Chapter 3
System Overview

Abstract Query processing in P2P networks poses inherent challenges and demands non-traditional techniques due to the distribution of content and the lack of global knowledge. Query routing over the P2P network is the key mechanism for efficient query processing. In the case of multidimensional data, designing multidimensional query routing is a non-trivial problem that should be carefully addressed. In this chapter, the challenges related to multidimensional query routing are briefly explained. Furthermore, the required routing information is described as well as the main components of distributed query processing.

3.1 Challenges of Distributed Multidimensional Query Processing

P2P multidimensional query processing refers to the execution of advanced query operators over multidimensional data stored in a highly distributed system. The main requirement for multidimensional query processing is the retrieval of the *exact* and *complete* result set. Exactness refers to retrieval of those points that belong to the query result, in contrast with approximate retrieval where points close or similar to the result points are retrieved. Completeness refers to the retrieval of all points that belong to the result set, in contrast to partial result retrieval that only retrieves a subset of the result. Equivalently, this means that although a query is processed in a distributed manner over a union of data sets $\cup S_i$, the result set should be the same as if the query had been executed on the dataset $S = \cup S_i$ in a centralized setting.

The objective of retrieval of exact and complete results greatly differentiates query processing from P2P information retrieval, where finding (a) similar results to the actual query results, or (b) an adequate portion of the result is considered acceptable. As a result, the aim of exactness and completeness in the result set produced by multidimensional query routing raises the complexity of query processing and poses new challenges for P2P retrieval.

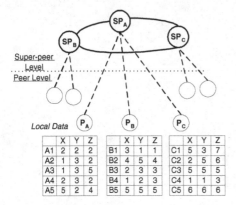

Fig. 3.1 Super-peer architecture

P_A	X	Y	Z
A1	2	2	2
A2	1	3	2
A3	1	3	5
A4	2	3	2
A5	5	2	4

P_B	X	Y	Z
B1	3	1	1
B2	4	5	4
B3	2	3	3
B4	1	2	3
B5	5	5	5

P_C	X	Y	Z
C1	5	3	7
C2	2	5	6
C3	5	5	5
C4	1	1	3
C5	6	6	6

In the rest of this book, it is assumed that the aim is to support exactness and completeness of results during distributed query processing.

3.2 Data Model and System Architecture

From the perspective of data management, the main focus of this book is on distributed management of multidimensional data. Commonly, in such application domains, there exists a set of nodes in the system that have special roles, due to enhanced capabilities, including high processing power, high storage capacity, network bandwidth, as well as due to other properties such as increased availability. To take advantage of such nodes, approaches will be presented that rely on a super-peer infrastructure. Hence, nodes with enhanced capabilities play the role of super-peers in our system. In addition, plain peers correspond to users who wish to participate and use the application, thus they register their machines to the P2P system. Peers are completely autonomous entities; for example, with respect to data storage each peer may autonomously store its own data. The only prerequisite for peer participation in the system is that peers should support some common functionality, which is required for indexing and query processing. This functionality can be in the form of a library of tools and operations (access methods, query processing algorithms, etc.) which is made available and is deployed on each peer at registration time.

More formally, each *peer* is denoted as P_i and let N_p represent the total number of peers in the P2P system. Each peer P_i holds n_i d-dimensional data points, denoted as a set S_i $(i = 1, \ldots, N_p)$. In consequence, the complete data set S stored at any time in the P2P system is the union of all peers' datasets S_i: $S = \cup S_i$, and its size n is equal to $\sum_{i=1}^{N_p} n_i$.

The overall architecture contains peers with special roles, called *super-peers*. Let N_{sp} denote the number of super-peers in the system, and SP_i to refer to a super-peer, with $i = 1, \ldots, N_{sp}$. Super-peers comprise only a small fraction of the peers in

Symbols	Description
S	Dataset
d	Data dimensionality
n	Cardinality of the dataset
S_i	Partition of dataset ($i = 1, \ldots, N_p$)
N_p	Number of peers
N_{sp}	Number of super-peers ($N_{sp} \ll N_p$)
DEG_p	Degree of simple peer
DEG_{sp}	Degree of super-peer ($DEG_{sp} < DEG_p$)

Table 3.1 Overview of basic symbols

the network, therefore $N_{sp} \ll N_p$. With respect to data storage, when a super-peer becomes part of the system initially, the super-peer may store its own data (in an analogous way to peers) or not (when the super-peer is used simply as infrastructure for the system). In this book, we assume that super-peers do not store their own data. Nevertheless, all techniques are also applicable also for the case where super-peers store their own data.

A super-peer maintains two sets of connections at logical level: (a) connections to peers, and (b) connections to other super-peers in the system. Super-peers accept a maximum number of DEG_p connections from peers that join the network directly and connect to one of the available super-peers. In addition, a super-peer is connected to a limited set of at most DEG_{sp} other super-peers ($DEG_{sp} < DEG_p$), which are called *neighbors* or *neighboring super-peers*.

Moreover, a multidimensional space D is used to represent the domain of data points. Thus, given a space D defined by a set of d dimensions $\{d_1, d_2, \ldots, d_d\}$ and a dataset S on D, a point $p \in S$ can be represented as $p = \{p_1, p_2, \ldots, p_d\}$, where p_i is the value of p on dimension d_i. For a reference to the basic symbols used in the remaining of this book, see Table 3.1.

3.3 Objectives and Motivation

3.3.1 Performance Objectives

A set of objectives related to performance issues is identified in the following, which eventually lead to the establishment of an efficient query processing mechanism. The aforementioned objective of exactness and completeness of the result also appears in this list abusively, even though it is not directly related with performance, due to its importance in the context of this book.

- **Contact only peers and super-peers that contribute to the query.** A basic prerequisite for P2P query processing is the creation of routing information for deliberate query routing, which is typically in the form of data summaries. The number of contacted peers and super-peers is influenced by the accuracy of the selected summary information. In general, in the absence of the detailed data, the summary information inherently involves some inaccuracy. The objective is to minimize the number of peers and super-peers that will be contacted in vain. Moreover, most of the algorithms presented in this book for distributed querying follow a thresholding scheme that allows pruning peers or super-peers based on already retrieved results.

- **Minimize the number of transferred data objects.** The employed thresholding scheme enables excluding objects that cannot appear in the query result set. In addition, peers and super-peers process the query locally and only points that belong to the local result set are transferred. Progressive merging at intermediate peers, where non-qualified data points as results are identified and removed, can be additionally employed to further reduce the amount of transferred data.
- **Minimize the processing time at each peer and super-peer.** This is achieved by utilizing efficient local access methods at each peer and super-peer, by means of adopting state-of-the-art indexing techniques proposed for centralized settings. As already mentioned, appropriate thresholds are used to query fewer peers and eagerly prune the search space.
- **Minimize the query response time.** This goal is naturally derived from the previous objectives, since minimizing all the above objectives has the consequence of minimizing the response time. It should be noted that some aims are contradictory and lead to interesting trade-offs. For example, merging of retrieved results at each intermediate super-peer leads to minimizing the transferred data, but increases the processing time at each super-peer. In this book, various methods are investigated for reducing the response time, such as the exploration of the advantages and disadvantages of different result merging strategies.
- **Exact and complete result set.** Apart from reducing the total response time, the combination of threshold usage with the summary information should additionally guarantee that no data object or (super-)peers that contribute to the result set is pruned by accident.

3.3.2 Inadequacy of Basic Approaches

Several well-known techniques for distributed query processing could potentially be applied in the case of multidimensional query processing and routing in a super-peer network. However, as will be shown presently, existing approaches present various shortcomings that make them inapplicable for the proposed setup.

Perhaps the most plain solution is to follow a data shipping approach, by having all peers send their local datasets to the querying peer, and then have the query processed locally over the complete dataset, identically with the case of centralized data storage. However, this approach is infeasible in a large-scale P2P system, because it is not scalable with network size. In addition, the required data transfer imposes high costs, both in terms of bandwidth consumption as well as total response time for producing the result of the query.

Basic search methods in P2P systems include variations of flooding and random walks. Such methods work acceptably well in particular cases; for example, when data is replicated to many peers in the system, or in special types of retrieval, such as document retrieval, where it suffices to retrieve only some relevant documents to the query. To ensure completeness of the result, these approaches would need to contact all peers in practice. As such, both approaches are not suitable for the retrieval of the complete query result set, because of the high incurred cost of search. In the case of super-peer networks, it is necessary to maintain information at super-peer level that describes the available data at peers, so that each super-peer can route the query only to those peers that store relevant data.

Routing indices [42] are routing structures that, given a query at a peer, identify the subset of neighbors that can produce results (if that direction was followed) ranked according to the number of query results. [42] studies the use of routing indices for document retrieval, and the goodness of a peer is reflected on the number of documents available through each neighbor. A peer receiving a query, first processes the query based on its local data, and when the retrieved documents are not enough, then the query is forwarded to the best neighbor based on the routing list. Alternatively, more than one neighbor can be queried simultaneously. Querying more than one peer in parallel reduces the response time, but increases the network traffic, since more messages are sent and more data objects are transferred to the querying peer. This approach assumes that there exists a stopping condition, i.e., the number of relevant documents to be retrieved, which can be easily evaluated at each peer, by counting the retrieved documents and attaching the number of already retrieved documents to the query.

3.3.3 Motivation for Summary Information

In contrast to the aforementioned case, in the case of complete retrieval, one major problem is the lack of a stopping condition that enables the determination of when the complete result set has actually been retrieved. Therefore, super-peers have to store multidimensional summary information to determine whether there exist more peers (or super-peers) that may contribute to the result set, before contacting them. This information serves as an aggregate summary of the available data objects, and is more coarse or abstract than the individual data items stored at the peers. However, in principle, data summaries may lead to overcount or undercount the number of returned results. Therefore, the employed summaries should have the salient property of guaranteeing that any peer that stores data which belongs to the result set will be contacted.

Another interesting issue related to data summaries is the level of detail. Coarse-grained data summaries reduce the storage and maintenance cost of routing information, at the expense of higher query processing cost. Instead, fine-grained data summaries support more efficient query processing, but include a higher cost of storage and maintenance. Clearly, the level of detail of the summarization leads to an interesting trade-off between maintenance and query processing cost. The proposed framework makes the implicit assumption that queries are much more frequent than updates. Thus, the main focus is on improving query performance, rather than minimizing the cost that incurs due to the creation and maintenance of the routing information. Motivated by this observation, a pre-processing step is employed for building routing information, in order to improve the performance of query processing. The reader may draw parallels to the case of index creation before processing queries over a table in a traditional relational DBMS.

3.4 Data Summaries and Indexing

In order to achieve the aforementioned objectives, we describe issues related both with centralized and distributed data management, as well as distributed query processing algorithms. We analyze a framework for multidimensional query processing and routing over a super-peer system that relies on a *three level data summarization scheme*:

1. Selection of appropriate data summaries at local peer level
2. Subsequent construction of summary information at any super-peer, describing the data available at peers connected to the super-peer.
3. Construction of routing information in the form of routing indices at super-peer level by exploiting super-peer data summaries, in order to exclude peers and super-peers that cannot contribute with any results to the given query.

The main goal is to design efficient distributed algorithms for query processing that take into account result sets from other peers to prune data or peers that cannot contribute to the result set. To this end, data summaries that describe the local data are needed. A peer needs to summarize its data and use centralized indexing techniques to support efficient query processing. A super-peer is required to build and maintain data of two categories: (a) summary data, collected from its peers and serving as an indexing mechanism over its peers' data, and (b) information about data available in the rest of the network, collected through its neighboring super-peers for routing queries to remote peers.

3.4.1 Peer Summaries

Prior to the actual query processing, a pre-processing or construction phase is required to create and distribute the routing information. In the pre-processing phase, each peer may create an index on its local data, in order to answer efficiently a given query. In turn, each super-peer collects some multidimensional information, such as representative data objects, from its associated peers. A crucial issue is what kind of multidimensional information each super-peer should collect. The main characteristic is that this information or summary should be able to define which peers have relevant data to the query and also have the power to prune as many peers as possible. In this book, different query operators and various types of summaries that should be used will be studied, in order to support the queries.

3.4.2 Super-peer Aggregated Summaries

Each super-peer collects the summaries of its associated peers. Then, each super-peer aggregates the summaries into a new set of summaries that describe in a more compact way the data available on its associated peers. In more details, a super-peer can use the compact summaries to discard summary data that do not correspond to results during query processing. Each super-peer may use any centralized indexing technique to efficiently query the collected peer summaries. Based on this information, each super-peer is able to answer efficiently a given query by selecting and querying the appropriate peers that are connected to it.

3.4.3 Routing Indices

In order to create routing summaries, each super-peer summarizes the aggregated summary data. The goal is to create a more coarse data summary, that leads to smaller construction and maintenance costs. The routing summaries describe in a more abstract way the data available on its associated peers. Moreover, the routing summaries may also summarize the data available through some of its neighboring super-peers. These summaries are distributed to the other super-peers through its neighbors. As a result, each super-peer is able to route efficiently a given query over the entire network.

3.4.4 Churn

An important and distinguishing issue in P2P networks is fault-tolerance. The super-peer architecture makes the system more resilient to failures compared to other P2P systems. Super-peers have stable roles, but in the extreme case that a super-peer fails, its peers can connect to another super-peer using the basic bootstrapping protocol. A peer failure may cause the responsible super-peer to update its aggregated summaries. Only if churn rate is high, these changes are propagated to other super-peers.

As already mentioned, a peer joins the network by contacting a super-peer using the bootstrapping protocol. The bootstrapping super-peer SP_B uses its routing sum-

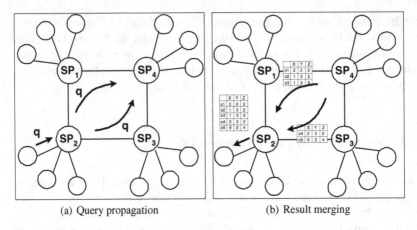

    (a) Query propagation   (b) Result merging

Fig. 3.2 Query processing over super-peer networks

mary to find the most relevant super-peer to the joining peer. This is equivalent to a similarity search over the super-peer network, and will be covered in Chapter 5. When the most relevant super-peer SP_r is discovered, the new peer joins SP_r. An interesting property of this approach is that joining peers become members of relevant super-peers, so it is expected as new peers join the system, that clustered data sets are gradually formed, with respect to the assigned super-peers. This is close to the notion of semantic overlay network (SON) construction [48], which is similar to having clustered data assignment to super-peers.

3.5 Framework

The main premise of efficient distributed algorithms for query processing is to exploit the routing indices to avoid querying peers and super-peers that cannot contribute to the result set. In addition, the result sets from already queried peers are taken into account to prune data or peers that cannot contribute to the result set. In the following, first a general description of query processing over super-peer networks is given, followed by an outline of query processing components.

3.5.1 Query Processing

Users (also modelled as peers) can submit queries that are propagated to an *initiator* super-peer, which then in turn routes the query selectively to those super-peers that can provide relevant data. The querying peer is denoted P_{init}, henceforth referred to as *initiator* of a query. Notice that even though the initiator can be a simple peer, P_{init} is used to refer to the super-peer responsible for the simple peer.

In super-peer architectures, queries are typically routed first in the super-peer backbone and afterwards, if necessary, they are distributed to the peers that are connected to the super-peers. In general, super-peers maintain information about the peers they have been assigned, so that at query time, they can process a query without having to contact all peers. Note that one important performance parameter is the super-peer topology which influences the routing performance. In this book, we assume that the pre-defined super-peer topology is assumed to be generic and not restricted to a particular form, and the focus is on the optimization of interactions among super-peers and peers.

In order to minimize the transferred data, the technique used is to locally evaluate as many parts of the query as possible. However, accurate and complete query computation over widely distributed data, demands that all data is taken into account, since even a single point neglected could affect the result set and prune out other already processed points.

Each super-peer SP_A that receives a query, routes the query selectively to those super-peers that can provide data belonging to the query result set. Then, super-peer SP_A evaluates a query by using its summary information about data stored at its peers, in order to find the peers that contribute to a query, and forwards the query only to the relevant peers. Each peer that receives the query, processes the query over its data and sends the results back to super-peer SP_A. This is the query propagation phase (Fig. 3.2(a)) that is followed by a merging phase (Fig. 3.2(b)). In the merging phase, each super-peer SP_A collects the results of the queried peers and neighboring super-peers. SP_A merges the local results into a new result set, in order to reduce the transferred data, or just sends the received results to the querying super-peer through the neighboring super-peers. Finally, the initiator gathers the objects and returns the result set to the querying peer.

3.5.2 Query Processing Components

Multidimensional query processing and routing over a super-peer network consists of the following *three main components*:

1. Local query processing at a peer based on its locally stored data (*local query processing*).
2. Local query processing at a super-peer based on aggregated summary data that is locally stored data (*peer selection* mechanism).
3. Query routing based on the routing indices (*super-peer selection* mechanism).

Local query processing requires the usage of efficient centralized techniques. In particular, for each query operator (range, nearest neighbor, top-k and skyline queries) supported by the proposed P2P system, efficient indexing methods of local data at peers are investigated. The goal of local query processing is to retrieve data points that are stored locally on this peer and belong to the result set. These data points are returned to the associated super-peer. The efficiency of local query processing depends on the centralized techniques that are employed. To improve the performance further, some information that describes data stored at other peers may be send to the peer together with the query in order to avoid retrieving data points that do not contribute to the result set. An additional advantage is that some of the local results are pruned early and are not send to the super-peer.

In contrast to local query processing, the goal of the peer selection mechanism is to select peers that store relevant data and not the data itself. The efficiency of the peer selection mechanism depends on the summary data stored on the super-peer as well as the performance of the query that retrieves the locally stored summary data. The peer selection mechanism should exclude peers and that cannot contribute with any results to the given query based on the summary information and on already retrieved results from a subset of peers.

The *super-peer selection* mechanism is responsible for exploiting the summary information that is exchanged between super-peers in order to identifying promising super-peers that may return relevant results to the query. Equally important is the issue of excluding (pruning) from further processing peers and super-peers that cannot contribute to final result set. Pruning is achieved based on the summary information exchanged between super-peers, but also by taking advantage of already retrieved results from a subset of peers. Thus, the goal of the super-peer selection mechanism is to identify the set of neighboring super-peers that should be queried.

The overall performance of query processing is influenced by all three components. The performance of each component in terms of response time influences the overall response time. More importantly, the efficiency of the peer and super-peer selection mechanism in terms of which peers or super-peers should be queried next influence the performance of query processing, since it affects the latency of the system and the amount of transferred data.

Chapter 4
Query Operators

Abstract Emerging applications over distributed, loosely coupled datasets require advanced query processing primitives that go beyond exact match queries. Such applications often need to handle multidimensional data, whether these dimensions are related to specific attributes of the data objects or are the result of advanced feature extraction algorithms. Querying multidimensional data is challenging even in a centralized domain. In this chapter we introduce fundamental query types that are commonly used in processing multidimensional data. We first discuss similarity search based on range queries and highlight their relation to nearest neighbor queries. Top-k queries, that rank data objects based on some scoring function are also discussed. We conclude our exposition with the recently introduced skyline query, as a generalization of ranking using many different, and often conflicting criteria. We discuss skyline computation over subspaces of the data domain and its relationship to top-k queries.

4.1 Multidimensional Data Model

During the last decades, an increasing number of applications, such as medical imaging, molecular biology, multimedia and scientific databases, have emerged where a large amount of high-dimensional data points needs to be processed. In addition to exact match queries, which have been largely supported both in centralized and distributed environments, emerging applications call for new, advanced query types that are particularly challenging in a distributed environment. For instance, queries like "which data objects are most similar to a query object" or "which data objects are the best trade-off between different object's features", require new query primitives, algorithmic and architectural solutions that are not available in existing systems.

As has been explained, we assume a data collection S of n objects represented as points in a d-dimensional (feature) space D characterized by dimensions $\{d_1, ..., d_d\}$. A data object is treated as a point p in D defined via a set of coordinate

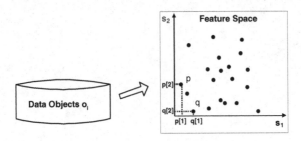

Fig. 4.1 Feature space

values in that space: $p = \{p_1, ..., p_n\}$. Each coordinate value p_i may represent an attribute of the object that is of interest to the application, or, it may be the result of a scoring function that evaluates certain features of the object [6]. In what follows, we assume that the points' coordinates are numerical non-negative values that depict certain features of database objects.

A two-dimensional example is shown in Fig. 4.1. In the figure, a database of objects o_j is depicted along with a representation of the objects as points in a two-dimensional space. The coordinates of each object are calculated via two scoring functions $s_1()$ and $s_2()$. Thus, object o_j is mapped to $p=(p_1,p_2)=(s_1(o_j),s_2(o_j))$. In our work, we do not distinguish between inherent attributes of the object and extracted features. We prefer to refer to its multidimensional representation and, we use the terms object and data point interchangeably.

Advanced query primitives have emerged in order to allow efficient processing of objects depicted in a high-dimensional space. Examples include similarity search based on range and nearest neighbor queries, top-k queries and the skyline operator. In the following, we shortly describe these query types.

4.2 Similarity Search in Metric Spaces

Several applications, such as multimedia databases [118], employ feature transformation, which projects important features or properties of data objects into a high-dimensional space. Subsequent processing in that space often requires support for similarity search in order to retrieve similar objects. Similarity search in metric spaces focuses on supporting queries, whose purpose is to retrieve objects which are similar to a query point, when a metric distance function *dist* is used to measure the objects' (dis)similarity.

More formally, a metric space is a pair $M = (D, dist)$, where D is a domain of feature values and *dist* is a distance function with the following properties:

1. $dist(p,q) \geq 0$ (non negativity),
2. $dist(p,q) = 0$ iff $p=q$ (identity of indiscernibles),
3. $dist(p,q) = dist(q,p)$ (symmetry),

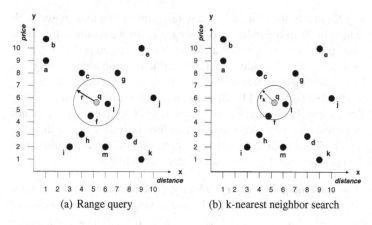

(a) Range query   (b) k-nearest neighbor search

Fig. 4.2 Similarity search examples

4. $dist(p,q) \le dist(p,o) + dist(o,q)$ (triangle inequality).

A smaller distance between two objects is used by the application to indicate higher degree of similarity among them. Similarity is symmetric, however, because of the triangle inequality, an object may be similar to two dissimilar objects.

Similarity search in metric spaces involves, two different types of queries, namely range and nearest neighbor queries.

4.2.1 Range Query

Range queries are specified by a query object q and a range (radius) value r. The result set of the query is defined to contain all the objects o from the dataset that have a distance to the query object q that is less than or equal to radius r:

Definition 4.1. Range query $R(q,r)$**:** Given a query object q and a radius r, a point $p \in S$ belongs to the result set R_q^r of the range query iff $dist(q,p) \le r$.

A range query $R(q,r)$ can be interpreted as "retrieve all objects that are within distance r to q". Fig. 4.2(a) depicts a range query defined by query point q and a radius r. The result set of this query contains data points f and l.

4.2.2 k-Nearest Neighbor Query

A drawback of range queries is that the cardinality of the result set is not known in advance, but can be anything between zero and the size of the database. Consequently, selection of an inappropriate value for the query range may lead to very few or too many query results. In the first case, a new range query has to be posed with a larger range, which leads to redundant processing cost. In the second case, more than necessary objects are retrieved, which again leads to increased processing cost. In practice, a good selection of a radius value r is difficult to obtain as it requires knowledge of the underlying distribution of data in the projected feature space.

The k-nearest neighbor (k-NN) [111, 70] query overcomes this problem by giving the user the ability to specify the size k of the answer set. This query type does not require a user to provide a query range and is therefore far easier to use than the similarity range query. The k-nearest neighbor query returns the k most similar (to a query point q) data points from the dataset and is defined as follows:

Definition 4.2. k-nearest neighbor query $NN_k(q)$**:** Given a query object q and a positive integer k, the result set NN_q^k of the k-nearest neighbor, is a set such that $NN_q^k \subseteq S$, $|NN_q^k| = k$ and $\forall u,v : u \in NN_q^k, v \in S - NN_q^k$ it holds that $dist(q,u) \leq dist(q,v)$.

In the definition we assumed that the dataset contains more than k points ($n = |S| \geq k$), which is typically the case. Otherwise, the result of the query is, trivially, the set of all objects in S. Moreover, the k data objects with the smallest distance may not be unique. When more than one objects have the same distance to the query point, one or more of them may be chosen randomly in order to produce a result set containing exactly k points.

Intuitively, a k-nearest neighbor query states "retrieve the k objects in S which are closest in distance to a given object". Fig. 4.2(b) illustrates the works of the nearest neighbor query. Given the query point q the figure depicts the results of the 2-nearest neighbor search (k=2), namely points f and l.

Given a query object q, a k-nearest neighbor query is equivalent to a range query specified by query point q and a radius equal to the distance of the k-th neighbor.

Observation 4.1 *Given a query object q, let r_k be the distance of the k-th nearest neighbor p, i.e., $r_k = dist(q,p)$ then $\forall r \in NN_q^k$ it holds that $r \in R_q^{r_k}$.*

Of course the range query may return a few additional objects whose distance from q is exactly r_k. Observation 4.1 expresses that any nearest-neighbor query can be transformed and have its results computed via a range query, if the distance to the k-th nearest neighbor was known a-priori.

4.3 Top-k Query

For decades, top-k queries were mainly studied in the information retrieval research field, aiming at ranking text documents according to some query terms, both efficiently and effectively. More recently [24, 78] the data management community has realized the benefits of top-k queries in database systems and several efficient algorithms for their evaluation have emerged. Top-k queries on multidimensional

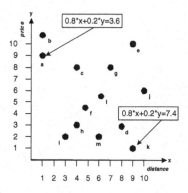

Fig. 4.3 Top-k example

datasets compute the k most interesting results with regards to a monotone score aggregation function, such as weighted aggregation, applied on the attribute values.

Definition 4.3. An aggregation function f is increasingly monotone, if $\forall p, p' \in S$ with $p_i \leq p_i', \forall i$, then $f(p) = f(p_1, ..., p_d) \leq f(p_1', ..., p_d') = f(p')$.

The property of increasing monotonicity means that whenever the score of all dimensions of the point p is at least as good as that of another point p', then we expect that the overall score of p is as least good as p'. The result of a top-k query is the ranked list of the k objects with lowest *score* values. As in the case of k nearest neighbor queries, when the database consists of fewer than k points, the result contains the whole dataset.

Definition 4.4. Top-k query: Given a positive integer k, the result set TOP_k of the top-k, is a set such that $TOP_k \subseteq S$, $|top_k| = k$ and $\forall u, v : u \in TOP_k, v \in S - TOP_k$ it holds that $f(u) \leq f(v)$, assuming that minimum values are preferable.

A special case of monotone functions is the weighted sum function, also called linear. Each feature (dimension) d_j has an associated query-dependent weight w_j indicating the dimension's relative importance for the query. The aggregated score for object p is defined as a weighted sum of the individual scores: $score(p) = \sum_{j=1}^{d} w_j \times p_j$, where $w_j \geq 0$ $(1 \leq j \leq d)$ and $\exists j$ such that $w_j > 0$. If some weights are set equal to zero, then a top-k query refers to only to a subset of the available features. The weights indicate the user's preferences and influence the ordering of the data objects and therefore the top-k result set. For example, consider the dataset depicted in Fig. 4.3. By assigning a high weight to values of dimension x (distance), point a is the top-1 object, while if a low weight is used, point k becomes the top-1 object.

A top-k query takes two parameters: a user specified monotone function f and the number of requested objects k. Notice that both the scoring function and the parameter k may differ for each query and we are interested in retrieving the k objects with the best (minimum) values of the scoring function. In the special case of

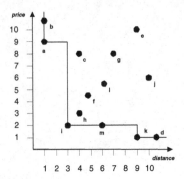

Fig. 4.4 Skyline example

the weighted sum, the user specifies the weighting of each feature, i.e., how important this feature is based on his preferences and therefore, a top-k query takes two
parameters: a d-dimensional vector $w = \{w_1, \ldots, w_d\}$ and the number of requested
objects k.

4.4 Skyline Operator

Skyline queries [21] have attracted much attention recently, since they help users
to make intelligent decisions over complex data, where many conflicting criteria
are considered. Let us assume for example a database containing information about
hotels. Each tuple of the database is represented as a point in a data space consisting
of numerous dimensions. In our example, the y-dimension represents the price of
a room, whereas the x-dimension captures the distance of the hotel to a point of
interest such as the beach (Fig. 4.4). According to the dominance definition, a hotel
dominates another hotel because it is cheaper and closer to the beach. Thus, the
skyline points, in the example points a, i and k, are the best possible trade-offs
between price and distance from the beach.

In the following, we define the skyline and subspace skyline queries and point
out their relation to top-k queries.

4.4.1 Skyline queries

Definition 4.5. Skyline: A point $p \in S$ is said to *dominate* another point $q \in S$, denoted as $p \prec q$, if (1) on every dimension $d_i \in D$, $p_i \leq q_i$; and (2) on at least one
dimension $d_j \in D$, $p_j < q_j$. The *skyline* is a set of points $SKY \subseteq S$ which are not
dominated by any other point. The points in SKY are called skyline points.

Without loss of generality, we assume that skylines are computed with respect to min conditions on all dimensions and that all values are non-negative.

The cardinality of the skyline set *SKY* depends on the data distribution, the dimensionality and the cardinality of the dataset. It has been shown [29, 61] that the expected number of skyline points is $\Theta(\ln^{d-1} n/(d-1)!)$ for a random dataset. The result suggests that the skyline cardinality increases with the dataset dimensionality. The intuition is that as the number of dimensions increases, it is more likely for any point p that there exists another point q, where p and q are better than each other in different subsets of dimensions. In other words, the probability of one point dominating another point in the full space is decreasing as the dimensionality increases. Therefore, the cardinality of the skyline set increases rapidly with the dimensionality of the dataset.

4.4.2 Subspace skyline queries

Applications often provide numerous candidate attributes that they can use for data analysis. In our running example, the hotel database could contain numerous other attributes, such as the number of rooms, the age of the hotel, the size of room, the star rating, etc. The notion of skyline can be extended to subspaces, where given a set of d-dimensional objects, a subspace skyline query only refers to a user-defined subset of attributes. Each non-empty subset U of D ($U \subseteq D$) is referred to as a *subspace* of D. The data space D is also referred as full space of the dataset S.

Definition 4.6. Subspace Skyline: A point $p \in S$ is said to *dominate* another point $q \in S$ on subspace $U \subseteq D$, denoted as $p \prec_U q$, if (1) on every dimension $d_i \in U$, $p_i \leq q_i$; and (2) on at least one dimension $d_j \in U$, $p_j < q_j$. The *skyline* of a subspace $U \subseteq D$ is a set of points $SKY_U \subseteq S$ which are not dominated by any other point on subspace U. The points in SKY_U are called *skyline points* on subspace U.

Consider for example the dataset depicted in Fig. 4.4. The skyline points are $SKY = \{a, i, k\}$, while for the (non-empty) subspace $U = \{x\}$ the skyline points on U are $SKY_U = \{a, b\}$. Notice that point b is a skyline point on the subspace $\{x\}$ but it is dominated by point a in the full space $\{x, y\}$.

Observation 4.2 *A skyline point $p \in SKY_U$ on a subspace $U \subseteq D$ is either a skyline point on D, or is dominated on D by another skyline point $q \in SKY_U$, for which $p_i = q_i$, $\forall i : d_i \in U$.*

4.4.3 Relation to top-k queries

Skyline queries relate to top-k queries, and can be used to discard points that cannot belong to the top-k result set.

Observation 4.3 *The top-1 object for any increasingly monotone aggregation function belongs to the skyline set.*

Proof: Consider a point q that does not belong to the skyline, but it is the top-1 for a query defined by an increasingly monotone function f. Then, there exists another point p that dominates q, i.e., on each dimension $d_i \in D$, $p_i \leq q_i$; and on at least one dimension $d_j \in D$, $p_j < q_j$, hence $f(p) < f(q)$, and since f is increasingly monotone this leads to a contradiction, because q is the top-1, i.e., $f(q) < f(p)$. Thus, the top-1 object for any increasingly monotone function belongs to the skyline.

For example, consider the dataset depicted in Fig. 4.3. By assigning a high weight to the score of attribute x, point a is the top-1 object, while if a low weight is used, point k becomes the top-1 object. Both a and k belong to the skyline set. This observation can be adopted for efficient top-k evaluation, by using the notion of the k-skyband operator [103]. The result set of any top-k query is a subset of the k'-skyband set, with $k \leq k'$.

Chapter 5
Similarity Search in Metric Spaces

Abstract Similarity search in metric spaces has attracted much attention recently due to numerous applications, including multimedia retrieval and scientific data management. Several centralized indexing methods have been proposed to support efficient similarity search in metric spaces. In this chapter, a distributed framework termed *SIMPEER* [49] is presented for efficient similarity search in P2P systems that extends the basic concepts of an efficient approach that has been previously proposed for centralized systems. SIMPEER dynamically clusters peer data, in order to build distributed routing information at super-peer level. The usage of carefully designed distributed data summaries guarantees that all similar objects to the query are retrieved, without necessarily flooding the network during query processing. With SIMPEER, the targeted query types (range and nearest neighbor queries) can be efficiently evaluated, thus reducing communication cost, network latency, bandwidth consumption and computational overhead at each individual peer.

5.1 Overview

This chapter addresses efficient processing of similarity queries in metric spaces, where data is horizontally distributed across peers in a super-peer network. The proposed framework termed SIMPEER builds and utilizes routing indices based on data summaries produced by means of distributed clustering.

In *SIMPEER* each peer maintains its own data objects, such as images or documents, which refer to a high-dimensional metric space and a distance function provides a measure of (dis)similarity. In order to make its data searchable by other peers, each peer first clusters its data using a standard clustering algorithm (such as K-Means), and then sends the cluster descriptions, typically consisting of the cluster centroid and radius, to its super-peer. Only the cluster descriptions as a summarization of the peers' data are published to the super-peer, while the original data is stored by the peer. The iDistance method [80, 144] is employed by the peer to index and provide access to its data, in order to efficiently answer similarity queries during local query processing.

Each super-peer maintains the cluster descriptions of its associated peers. In order to keep the information in a manageable size, each super-peer applies a clustering algorithm on the cluster descriptions of its peers, which results in a new set of cluster descriptions, also referred to as *hyper-clusters*, which summarize the data objects of all peers connected to the super-peer. The super-peer keeps a list of the hyper-clusters in main memory and stores in a B^+-tree the peers' clusters using an extension of the iDistance technique that is capable to handle cluster descriptions instead of data points. This extension, introduced in the next section, enables efficient similarity search over the cluster descriptions, so that the query is posed only to peers having data that may appear in the result set.

The remaining challenge is to answer such queries over the entire super-peer network. Instead of flooding queries at super-peer level, we build routing indices based on the hyper-cluster descriptions that enable selective query routing only to super-peers that may actually be responsible of peers with relevant results. The routing index construction is based on communicating the hyper-cluster descriptions and it is described in detail in Section 5.3. The number of collected hyper-clusters can be potentially large, therefore the super-peer applies a clustering algorithm that results in a set of *routing clusters*, that constitute a summary of the hyper-cluster information. In a completely analogous way to the indexing technique of the peers' clusters, the super-peer uses the proposed extension of iDistance to store the hyper-cluster information, this time maintaining in main memory only the routing clusters.

To summarize, *SIMPEER* utilizes a three-level clustering scheme:

- Each peer clusters its own data. The resulting clusters are used to index local points using iDistance.
- A super-peer receives cluster descriptions from its peers and computes the hyper-clusters using our extension of iDistance. Hyper-clusters are used by a super-peer to decide which of its peers should process a query.
- Hyper-clusters are communicated among super-peers and are further summarized, in order to build a set of routing clusters. These are maintained at super-peer level and they are used for routing a query across the super-peer network.

This chapter focuses on two query types that are frequently used in applications that support similarity search; range and nearest neighbor queries. The constructed routing indices based on cluster summarization support efficient query processing, in terms of local computation costs, communication costs and overall response time, for both range and nearest neighbor queries. A query may be posed at any peer and is propagated to the associated super-peer, which becomes responsible for the query processing and finally returns the result set to the querying peer.

Given a range query $R(q, r)$, each super-peer SP_A that receives the query uses its routing clusters to forward the query to the neighboring super-peers which have (either locally or in their routing indices) clusters that intersect with the query. Thereafter, SP_A forwards the range query only to its own peers that have clusters intersecting with the query based on the hyper-clusters, or in other words to peers that hold data that may appear in the result set and should therefore be examined. Finally, SP_A collects the results of its associated peers and the queried neighboring super-peers and sends the result set back to the super-peer (or peer in the case of the initiator) from which SP_A received the query.

To process nearest neighbor queries, the initiator super-peer is responsible to map this query to a range query and then propagate it to the neighboring super-peers based on its routing index. One of the arising challenges is the transformation of a nearest neighbor query to a range query. A k-NN query $NN_k(q)$ is equivalent to a range query $R(q, r_k)$ where r_k is the distance of the k-th nearest neighbor from the point q. The main problem is that the distance r_k is not known a priori. Therefore, a heuristic is required to estimate the distance of the k-th nearest neighbor. In SIMPEER, two alternatives are employed to estimate an appropriate range over the distributed data, based on histograms that capture distance distributions within clusters. Details on nearest neighbor search in SIMPEER can be found in [49], while this chapter focuses primarily on processing of range queries.

The remaining of this chapter is organized as follows: in Section 5.2, local query processing is described. Then, in Section 5.3, the employed routing summaries are described in detail. Section 5.4 presents the query routing and processing mechanisms established in SIMPEER. Alternative existing approaches are mentioned in Section 5.5, while Section 5.6 summarizes the main points of the chapter.

5.2 Local Data Summaries and Query Processing

5.2.1 Peer Local Data Summaries

Each peer is responsible for its own data, which is organized and stored based on the iDistance concept. First, the peer applies a clustering algorithm on its local data. Even though the choice of the algorithm influences the overall performance of the system, each peer may choose any clustering algorithm. The clustering algorithm leads to a set of k_p clusters $LC_p = \{C_i : (K_i, r_i) | 1 \leq i \leq k_p\}$. Each cluster C_i is described by a cluster centroid K_i and a radius r_i, which is the distance of the farthest point of the cluster to the centroid.

Each data object is assigned to the nearest cluster C_i and it is mapped to a one-dimensional value following the same mapping as iDistance. The iDistance values of the data objects are indexed in an ordinary B^+-tree, while the list of the clusters LC_p, with the centroids and the radii of the clusters, is kept in main memory.

iDistance [80, 144] is an indexing method proposed for high-dimensional similarity search in centralized databases. The main idea is to partition the data space into n clusters and select a reference point K_i for each cluster C_i. Then, each data object is assigned a one-dimensional iDistance value according to the distance to its cluster's reference object. Having a constant c to separate individual clusters, the iDistance value for an object $x \in C_i$ is

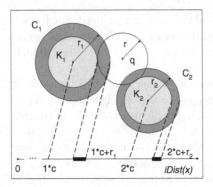

Fig. 5.1 *iDistance* mapping example

$$iDist(x) = i * c + dist(K_i, x)$$

Expecting that c is large enough, all objects in cluster i are mapped to the interval $[i * c, (i+1) * c]$, as shown in Fig. 5.1.

The actual data objects are stored in a B^+-tree using the iDistance values as keys. Additionally, the cluster centers K_i and the cluster radius r_i are kept in a main memory list. In this way, the problem of similarity search is transformed to an interval search problem. For a range query $R(q, r)$, for each cluster C_i that satisfies the inequality $dist(K_i, q) - r \leq r_i$[1], the data elements that are assigned to the cluster C_i and their iDistance values belonging to the interval $[i * c + dist(K_i, q) - r, i * c + dist(K_i, q) + r]$ are retrieved. For these points p_i the actual distance to the query point is evaluated and thereafter, if the inequality $dist(p_i, q) \leq r$ holds, p_i is added to the result set. In [144] an algorithm for nearest neighbor search was proposed based on repetitive range queries with growing radius.

5.2.2 Local Query Processing

The basic query processing functionality on a peer consists of an algorithm for processing of range queries over its local data. As stated in [80, 144], the algorithm examines each cluster in the list LC_p and searches separately those clusters that possibly contain objects matching the query. Algorithm 5.1 provides the pseudocode of how range query processing on a peer is performed. Practically, for each peer cluster $C_i \in LC_p$, the algorithm tests if the query intersects the cluster area (line 4). Thus, if a cluster C_i satisfies the inequality $dist(K_i, q) - r \leq r_i$, an interval search $[dist(K_i, q) + i * c - r, dist(K_i, q) + i * c + r]$ is posed on the B^+-tree. This iDistance interval corresponds to the cluster area that should be scanned in order to find all relevant objects. After these objects are retrieved, a refinement step is required, due

---

[1] Henceforth mentioned as intersection of the range query $R(q, r)$ and the cluster C_i.

Algorithm 5.1 Peer query processing

 1: **Input:** (q, r)
 2: **Output:** Result set S
 3: **for** $C_i \in \{LC_p\}$ **do**
 4:     **if** $(d(K_i, q) - r \leq r_i)$ **then**
 5:         $cursor \leftarrow B^+ tree\_range\_query[dist(K_i, q) + i * c - r, dist(K_i, q) + i * c + r]$
 6:         **while** $(candidate = has\_next(cursor))$ **do**
 7:             **if** $(dist(candidate, q) \leq r)$ **then**
 8:                 $S \leftarrow S \cup \{candidate\}$
 9:             **end if**
10:         **end while**
11:     **end if**
12: **end for**
13: **return** S

---

to the lossy mapping of iDistance, which maps different equidistant points from K_i to the same one dimensional value. In the refinement step, each object's distance to q is computed and if it is smaller than r (line 7), the object is added to the result set S (line 8). For example, in Fig. 5.1 the range query intersects with both clusters C_1, C_2. According to the iDistance values all objects enclosed in the dark grey shadowed area are retrieved and examined whether they belong to the result set.

5.2.3 Super-peer Local Data Summaries

A super-peer SP_A processes a range query by using its peers' cluster descriptions. Therefore, a super-peer determines the clusters and, consequently, also the peers that intersect with the range query, while the actual data is accessed directly from peers during query processing. Since the number of clusters increases rapidly according to the number of connected peers, in order to reduce the number of distance computations and intersection calculations and provide efficient query processing, the iDistance concept is followed. SP_A applies a clustering algorithm on the cluster descriptions and – in a similar way to iDistance – maps high-dimensional points to one-dimensional values. The cluster descriptions (having high-dimensional cluster centers) are mapped to one-dimensional values, in such a way that range and k-NN queries can be mapped into an interval search. In the following, the extension of the iDistance mapping for clusters is described.

5.2.3.1 One-dimensional Mapping of Clusters

A super-peer SP_A collects the cluster descriptions from its associated peers $LC_{sp} = \{(K_1, r_1), ..., (K_{n_{sp}}, r_{n_{sp}})\}$, where n_{sp} is the total number of clusters. For the sake of simplicity, we assume that $n_{sp} = k_p * DEG_p$, i.e., all peers have the same number of clusters k_p. Following the iDistance concept, SP_A applies a clustering algorithm on the list LC_{sp} which results in a list of clusters (called hyper-clusters) $LHC_{sp}=\{HC_i : (O_i, r_i')|1 \leq i \leq k_{sp}\}$, where k_{sp} the number of hyper-clusters, O_i the hyper-cluster center and r_i' the hyper-cluster radius, which is the distance of the farthest point of all clusters assigned to the hyper-cluster, to the centroid.

Each cluster C_j is mapped to a one-dimensional value based on the nearest hyper-cluster center O_i using formula:

$$key_j = i * c + [dist(O_i, K_j) + r_j]$$

which practically maps the farthest point of a cluster C_j based on the nearest reference point O_i. Similarly to iDistance, the one-dimensional values are indexed using a B^+-tree. In more detail, the B^+-tree entry e_j consists of the cluster's center K_j, its radius r_j and the distance d_j to its nearest reference point:

$$e_j : (key_j, K_j, r_j, d_j, IP_j)$$

Additionally, in the B^+-tree entry, the IP address of the peer is stored, in order to be able to propagate the query to those peers that have clusters that intersect with the query.

Furthermore, for each hyper-cluster HC_i, except from the radius r_i', we also keep a lower bound ($dist\_min_i$) of all cluster distances. The distance $dist\_min_i$ is the distance of the nearest point of all clusters C_j to O_i. These two distances practically define the effective data region of reference point O_i, or in other words, the region where all points of all clusters C_j belong to.

5.3 Routing Summaries

As regards routing index construction, each super-peer builds a variant of routing indices, in order to efficiently route queries to the appropriate neighboring super-peers. The routing information consists of assembled hyper-clusters HC_i of other super-peers. In more detail, for each neighboring super-peer a list of hyper-clusters is maintained, corresponding to hyper-clusters that are reachable through this particular neighboring super-peer. During query routing, the routing indices are used to prune neighboring super-peers, thus inducing query processing only on those super-peers that can contribute to the final result set. More formally, given a query $R(q, r)$ and a set of hyper-clusters $HC_i:(O_i, r_i)$, a neighboring super-peer is pruned if for all of its hyper-clusters HC_i it holds:

$$dist(O_i, q) > r + r_i$$

Each super-peer SP_A broadcasts its hyper-clusters using *create* messages in the super-peer network. Then, each recipient super-peer SP_r reached by *create* messages, assembles the hyper-clusters of other super-peers. Even though this broadcasting phase can be costly, especially for very large super-peer networks, it should be emphasized that this cost 1) is paid only once at construction time, 2) depends mainly on the number of super-peers N_{sp} and hyper-clusters per super-peer k_{sp} and not on the cardinality n of the data set, as in the case of structured P2P networks, and 3) can be tolerated for the network sizes of currently deployed super-peer networks.

Since the list of assembled hyper-clusters at SP_r may be potentially big to maintain in main memory, SP_r runs a clustering algorithm, which results in a set of *routing clusters (RC)*. Then, SP_r indexes the hyper-clusters, in a completely analogous manner as it clustered its peers clusters into hyper-clusters. The only difference is that for each routing cluster, the identifier of the neighboring super-peer, from which the owner of the hyper-cluster is accessible, is additionally stored.

Summarizing, a super-peer SP_A uses the extension of the iDistance in two ways. First, it clusters its peers' clusters $\{LC_{p_i}|1 \leq i \leq DEG_p\}$ into hyper-clusters HC_i and indexes this information in a B^+-tree. SP_A also clusters the hyper-clusters $\{LHC_i|1 \leq i \leq N_{sp} - 1\}$ collected from other super-peers, resulting in a list of routing clusters LRC_A. These are then used to index LHC_i in a separate B^+-tree using the one-dimensional mapping of iDistance, analogously to the technique employed for its peers' clusters.

Maintenance of routing indices is straightforward and can be accomplished using the well-established techniques described in [42]. In practice, when a super-peer detects a significant change in one of its hyper-clusters, it decides to broadcast this modification in a similar way to the construction phase. A modification can be due to peer data updates that altered the radius of a peer cluster, and eventually its hyper-cluster. It can also be the result of churn (peer joins or failures), that alter the radius of a hyper-cluster.

5.4 Query Routing and Processing

In this section, we provide an algorithm that retrieves all clusters that intersect with a given range query $R(q,r)$. In order to retrieve all clusters that belong to a hyper-cluster HC_i an interval search $[i * c + dist\_min_i, i * c + r'_i]$ on the iDistance values is posed, since the region $[dist\_min_i, r'_i]$ contains all clusters assigned to the hyper-cluster HC_i. In the following, we denote with dis the distance of O_i to q. The goal of our search algorithm is to filter out clusters that do not intersect with the query, based on the iDistance values. Since the points are mapped to one-dimensional values with respect to the farthest points of each cluster, searching all indexed points until r'_i cannot be avoided. This is clearly depicted in Fig. 5.2 by means of an example. The hyper-cluster radius r'_i is defined by the farthest point of the cluster C_1, whereas $dist\_min_i$ is defined by cluster C_5. The query intersects with C_1 that is mapped to an iDistance value based on the r'_i distance. In other words, it is not enough to search until $dis + r$, since some farthest points of intersecting clusters may be overlooked. The starting point of the interval search is the iDistance value corresponding to $max(dis - r, dist\_min_i)$. For the query $R(q,r)$, in our example (Fig. 5.2), the search scans the interval $[i * c + dis - r, i * c + r'_i]$.

Algorithm 5.2 describes the range query search algorithm performed by super-peer. Range query search takes as input a query point q and a radius r. The range search algorithm essentially consists of three steps: 1) it checks whether the hyper-cluster HC_i can provide relevant results (line 6), 2) (if so) it locates a starting point,

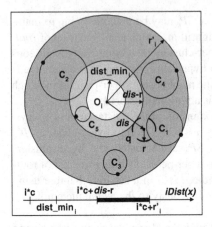

Fig. 5.2 Search interval based on range query $R(q,r)$

Algorithm 5.2 Range query search

1: **Input:** (q,r)
2: **Output:** Result set S
3: **for** $O_i \in \{LHC\}$ **do**
4:     $dis \leftarrow dist(O_i, q)$
5:     $lower \leftarrow max(dis - r, dist\_min_i)$
6:     **if** $(dis - r \leq r_i')$ **and** $(dis + r \geq dist\_min_i)$ **then**
7:         $lnode \leftarrow LocateLeaf(btree, i * c + lower)$
8:         $Outward(lnode, i * c + r_i', dis)$
9:     **end if**
10: **end for**
11: **return** S

---

denoted as $lower = max(dis - r, dist\_min_i)$ (line 5), for starting an interval search
on the B^+-tree (line 7), and 3) scans the interval until r_i' (line 8). *LocateLeaf()* im-
plements a standard search for some input value on a B^+-tree and returns the leaf
node corresponding to the input value. Notice that the same notation is used as in
the original iDistance publication [144].

Outward (Algorithm 5.3) takes as input the leaf node from which the search
starts, the high-end value for searching and the distance *dis* of the query point to
the hyper-cluster. Each entry e_i, i.e., cluster, stored in node is checked to ensure that
it is within distance $r + r_i$ to the query (line 4). The objective of the range query
algorithm is to reduce not only the number of accessed nodes, but also the number
of distance computations needed for query processing, which may be costly for
complex distance functions in metric spaces. For this purpose, all the information
concerning distances stored in the nodes of the B^+-tree is used to effectively apply
the triangle inequality. The value $|dis - d_i|$ (line 3) is a lower bound on the distance
of $dist(K_i, q)$. The inequality in line 3 reduces the number of distance computations
during query processing. If necessary, the algorithm initiates a search to the next
leaf (line 9-10) of the B^+-tree.

Algorithm 5.3 Outward

1: **Input:** $(node, ivalue, dis)$
   /* $\{E\}$ is the set of entries in $node$ */
2: **for** $(e_i : (key_i, K_i, r_i, d_i, IP_i) \in \{E\}; key_i < ivalue)$ **do**
3:     **if** $(|dis - d_i| \le r + r_i)$ **then**
4:         **if** $(dist(K_i, q) \le r + r_i)$ **then**
5:             $S \leftarrow S \cup e_i$
6:         **end if**
7:     **end if**
8: **end for**
9: **if** $(e_{last}.key < ivalue)$ **then**
10:    Outward($node.rightnode, ivalue$)
11: **end if**
12: **return**

k-NN Search. In a similar way to iDistance, peers are capable to process nearest neighbor queries locally. According to [80, 144], a peer would process the query by executing a range query with an initial range. If fewer than k data objects are retrieved, the radius of the range query is increased repeatedly, until k data objects are retrieved. Since our application area is a P2P environment, a strategy that uses a small radius and increments it until k objects are retrieved would cause more than one round-trips over the network, which is quite costly. To avoid this pitfall, in *SIMPEER* the super-peers maintain a set of statistics to estimate the distance of the k-th object from the query point and avoid the execution of multiple range queries. The specific details about the type of statistics as well as the algorithm used to perform nearest neighbor search can be found in [49].

5.5 Other Existing Approaches

The design of efficient indexing and query processing techniques over high-dimensional datasets has attracted the attention of many researchers, due to database applications, such as multimedia retrieval and time series matching, which deal with high-dimensional data representations. In such applications, an important issue is to retrieve similar objects to a given query object. Typical operations for similarity search include range and k-nearest neighbor (k-NN) queries. Different approaches have been developed for the problem of k-NN query processing, when k is known in advance [111], whereas other approaches deal with the problem of incremental nearest neighbor search [69, 70], where k is unknown and the query processing stops on demand. It should be noted that, in some applications, an algorithm that returns approximate results is sufficient, therefore fast algorithms have been developed for retrieving approximate nearest neighbor, however in the following the focus is on algorithms that retrieve the exact result set.

5.5.1 Centralized Algorithms

Different algorithms have been proposed for supporting k-NN queries based on multidimensional index-structures, like R-trees [65]. Multidimensional indexing has been extensively studied in the past. The interested reader should refer to [57] for a survey of various techniques. In [111], an efficient branch-and-bound search algorithm for processing exact k-NN queries based on R-trees is presented and several metrics for ordering and pruning the search tree are introduced. In [70], an incremental nearest neighbor algorithm is presented that is applicable to the R-tree. In [127, 15], cost models for nearest neighbor search in a high-dimensional data space are presented. In general, the performance of k-NN algorithms based on multi-dimensional indexes highly depends on the quality of the underlying index.

To alleviate the problem of poor performance of the indexes in high dimensions, k-NN algorithms were proposed that apply dimensionality reduction techniques, such as [56, 19]. The performance of multi-step algorithms for k-NN queries, i.e., algorithms that use an index on a data space of reduced dimensionality, were analyzed in [119]. In [25], an approach called Local Dimensionality Reduction (LDR) is presented that also exploits multiple indexes, one for each cluster, unlike other dimensionality reduction methods, that aim to reduce the dimensionality over the entire dataset and are insufficient for datasets that are not globally correlated. In [143], an approach (iMinMax) where the high-dimensional points are mapped into a one-dimensional space was proposed, by mapping high dimensional points to single-dimensional values determined by their maximum or minimum values among all dimensions. In [16], the effect of dimensionality on the nearest neighbor problem is explored. The authors show that as dimensionality increases, the distance of any data point to its nearest data point approaches the distance to its farthest data point. Therefore, in [68], an approach is proposed that does not treat all dimensions equally, but uses a quality criterion to select relevant dimensions with respect to the given query.

The aforementioned algorithms assume that data objects are represented in a d-dimensional Euclidean space and therefore distances can be computed between any two points in the data space. However, in several applications, similarity search algorithms are required that operate on metric spaces, where the exact distance function may be unknown. In this case, only a distance matrix is available that captures the value of the distance between any two existing data objects. Algorithms for similarity search in metric spaces only use of the properties of a metric distance function (nonnegativity, symmetry, and the triangle inequality), and can in principle operate without any knowledge of how the distances between objects have been computed.

Similarity search in metric spaces has attracted much attention recently, mainly due to important applications, such as image retrieval. Several indexing approaches [37, 144, 80] have been proposed to handle efficient similarity search in metric spaces for centralized settings. A dynamic balanced index structure, called M-tree, for similarity search in metric spaces, was presented in [37]. A branch-and-bound technique is proposed to efficiently retrieve the k-nearest neighbor data objects. In [38], a cost model for querying the M-tree index is developed that uses the distance distribution of objects. iDistance [144, 80] is an index method for similarity search. It partitions the data space into a set of clusters and selects a reference point for each cluster. Then, each data object is assigned a one-dimensional iDistance value according to the distance to its cluster's reference object. The actual data objects are stored in a B^+-tree using the iDistance values as keys. Additionally, the cluster centers and the cluster radii are kept in a main memory list. In this way, the problem of similarity search is transformed to an one-dimensional interval search problem. Interesting surveys of similarity query processing in metric spaces can be found in [71, 32].

5.5.2 Distributed Algorithms

Similarity search in P2P systems has attracted a lot of attention recently, however most existing approaches focus mainly on structured P2P systems or on building a network topology that groups together peers with similar content.

Recently, MCAN [55] and M-Chord [100] were proposed to handle similarity search in metric spaces, employing a structured P2P network. Both approaches focus on parallelism for query execution, motivated by the fact that in real-life applications, a complex distance function is typically expensive to compute. MCAN uses a pivot-based technique that maps data objects to an N-dimensional vector space, while M-Chord uses the principle of iDistance [80] to map objects into one-dimensional values. Afterwards, both approaches distribute the mapped objects over an underlying structured P2P system, namely CAN [106] and Chord [121] respectively. Queries are transformed into a series of interval queries, executed on the corresponding structured P2P system. It should be noted that data preprocessing (clustering and mapping) is done in a centralized fashion, and only then data is assigned to peers. Relevant to this work, Batko et al. [12] present a comparative experimental evaluation of four distributed similarity search techniques (VPT*, GHT*, MCAN, M-Chord). VPT* and GHT* [11] are two distributed metric index structures where the dataset is distributed among peers participating in the network. A conclusion that is drawn is that for single query execution the GHT* is the most suitable data structure, whereas for multiple queries M-Chord performs best.

Recent works aim to process similarity search in P2P systems by building a *suitable overlay topology*. A general solution for P2P similarity search for vector data is proposed in [8], named SWAM. Unlike structured P2P systems, peers autonomously store their data, and efficient searching is based on building an overlay topology that brings nodes with similar content together. However, SWAM is not designed for metric spaces. A P2P framework for multi-dimensional indexing based on a tree structured overlay is proposed in [82]. In [45], Datta et al. study range queries over trie-structured overlays. LSH forest [13] stores documents in the overlay network using a locality-sensitive hash function to index high-dimensional data for answering approximate similarity queries.

Most approaches that address *range query processing* in P2P systems rely on space partitioning and assignment of specific space regions to certain peers. A load-balancing system for range queries that extends Skip Graphs is presented in [120]. The use of Skip Graphs for range query processing has also been proposed in [58]. Several P2P range index structures have been proposed, such as Mercury [17], P-tree [41], BATON [81]. A variant of structured P2P for range queries that aims at exploiting peer heterogeneity is presented in [101]. In [92], the authors propose NR-tree, a P2P adaptation of the R*-tree, for querying spatial data. Recently, in [84], routing indices stored at each peer are used for P2P similarity search. Their approach relies on a freezing technique, i.e., some queries are paused and can be answered by streaming results of other queries. Maintenance techniques for routing indices have been studied in [74].

There also exists some work on P2P similarity search that focuses on *caching* [64, 115] or *replication* [18]. In [64] an architecture is presented for a data sharing peer-to-peer system, where the data is shared in the form of database relations. Approximate range queries are processed based on a range-caching scheme developed on a DHT to support range queries. Another approach, called MAPLE, that uses query result caching is presented in [87]. MAPLE is designed for the efficient sharing of query results cached in the local storage of mobile peers, in order to provide efficient location-dependent nearest neighbor search on each host.

Moreover, *P2P similarity search for document retrieval* has been studied in [114]. A structured P2P network is used that introduces a generic content-based similarity search scheme, based on the LSI (latent semantic indexing) model for document retrieval. At an initial step, a sample-based LSI computation at a central peer is proposed. Thereafter, the LSI model is broadcast to all peers and their document representations are locally computed. Based on a set of preselected reference documents, the document's similarity to the reference set is calculated and used to construct a key for publishing documents in Chord. A quite similar approach is the system called pSearch [123]. This work focuses mostly on finding similar documents over a P2P network and cannot be easily applied to metric spaces in general.

Various extensions of SIMPEER have recently appeared addressing different application requirements. In [50], range queries with recall guarantees in metric spaces have been studied. Improving the performance of multidimensional routing indices built over a super-peer network has been the focus of [51, 52].

5.6 Summary

In this chapter, we discussed efficient similarity search in P2P networks. It is noteworthy, that the presented framework is designed for metric spaces, thereby supporting non-Euclidean distance functions, thus becoming suitable for applications like distributed image retrieval and document retrieval in distributed digital libraries. The framework is based on a novel three-level clustering scheme and utilizes a set of distributed data summaries. Algorithms for efficient distributed range query processing are provided, while k-nearest neighbor queries are mapped into range queries based on an estimated radius.

Chapter 6
Subspace Skyline Queries

Abstract Skyline queries help users make intelligent decisions over complex data, when different and often conflicting criteria are considered. Such queries return a set of data points that are not dominated by any other point on all dimensions. Skyline queries have been studied in centralized systems and more recently in distributed environments, such as web information systems and peer-to-peer (P2P) networks. Skyline query processing in P2P networks poses inherent challenges and demands non-traditional techniques due to the distribution of content and the lack of global knowledge. Distributed skyline processing should minimize transfer of globally dominated data points. Moreover, peers that do not store any global skyline points, ideally should not be contacted at all. Detecting domination between points that are stored by different peers is challenging and requires usage of summary information that describes the data stored locally and across the network. In this section, we describe in details a distributed framework for subspace skyline processing, called SKYPEER+ [130]. SKYPEER+ reduces both computational time and volume of transmitted data due to (i) efficient routing of skyline queries over the super-peer network, and (ii) an effective thresholding and indexing scheme for discarding dominated points.

6.1 Overview

In this chapter, we explore the implications of processing and routing subspace skyline queries in large scale P2P networks. Given a subspace skyline query defined on a user-specified subspace U, a key observation is that a point p belongs to the global subspace skyline denoted as SKY_U, only if there exists a data partition S_i where p is a local skyline in U. Thus, only local skyline points have to be taken into account during a distributed skyline evaluation. In order to avoid contacting all peers during query processing, in a pre-processing phase each super-peer gathers some summary information that compactly describes the data stored at each peer.

In more details, each peer calculates a subset of the original dataset, called *extended skyline* (ext-skyline), that contains all the skyline points for any subspace U defined in a user query. In order to minimize unnecessary data transfer, the main premise is to evaluate as many parts of the query as possible locally, at each super-peer. However, accurate skyline computation over widely distributed data, demands that all relevant data is taken into account, since even a single point neglected could be part of the skyline and, thus, prune out other points already processed. As will be explained, routing indices that rely on the extended skyline sets are sufficient to determine if a peer P_i can contribute to the result set of a given skyline query on an arbitrary subspace $U \subseteq D$.

Moreover, subspace skyline points already computed at another super-peer may dominate –and thus prune – points of the current super-peer. Therefore, a threshold value can be defined based on already computed subspace skyline points, and this threshold is attached to the query before it is propagated further in the network. The threshold value can be updated at any super-peer in the network during query processing based on local query results. This way, the super-peers are able to detect and discard dominated data points across different peers. By using routing indices that compactly summarize data across super-peers, the thresholding scheme can be further improved and the transferred data can be reduced drastically. Each super-peer clusters the data of its peers in order to provide a succinct summary and make this data searchable by other super-peers. The cluster descriptions are published in the network helping build routing indices. This process facilitates an effective indexing and routing mechanism for subspace skyline queries over a super-peer network.

The rest of this chapter is organized as follows: In Section 6.2, we introduce the extended skyline and an intuitive one-dimensional mapping of the data that enables threshold-based pruning of data points during skyline evaluation based on already computed skylines. Section 6.3 describes the local data summaries, discusses indexing of the local ext-skylines and presents an efficient algorithm for computing local subspace skyline points in response to a user query. In Section 6.4, we introduce routing indices, while in Section 6.5 we present efficient algorithms for the computation and routing of user queries in the network. In Section 6.6, relevant approaches to distributed skyline processing are described. Finally, in Section 6.7, we summarize the chapter.

6.2 Dominance Relationships in Arbitrary Subspaces

6.2.1 Extended Skyline

Skyline query processing in P2P networks poses inherent challenges and demands non-traditional techniques due to the distribution of content and the lack of global knowledge. In order to avoid unnecessary transmission of data in the P2P network,

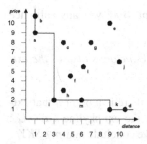

Fig. 6.1 Extended skyline example

in a pre-processing step we calculate a subset of the original dataset that contains all the skyline points for any subspace.

In this chapter, we follow a slightly different notation because of the existence of subspaces. Given a space D defined by a set of d dimensions $\{d_1, d_2, .., d_d\}$ and a dataset S on D, a point $p \in S$ can be represented as $p = \{p[1], p[2], ..., p[d]\}$ where p[i], is a value on dimension d_i.

Consider for example the dataset depicted in Fig. 6.1. The skyline points are $SKY = \{a, i, k\}$, while for the subspace $U = \{y\}$ the skyline points on U are $SKY_U = \{k, d\}$. Notice that point d is a skyline point on the subspace $\{y\}$ but it is dominated by point k in the full-space $\{x, y\}$. As is denoted in the following two observations [145], the set of global skyline points does not contain all the skyline points for any subspace.

Observation 6.1 *Given a set S of data points on dimension set D, for two subsets U and V of D ($U, V \subseteq D$), where $U \subseteq V$, there is no containment relationship between SKY_U and SKY_V.*

Observation 6.2 *Assume a set S of data points on dimension set D and two subsets U and V of D ($U, V \subseteq D$) such that $U \subset V$. Each skyline point q in SKY_U on dimension set V is either dominated by another skyline point p in SKY_U; or a skyline point in SKY_V.*

Based on the above observations, a skyline point q in SKY_U is either a skyline point in SKY_V (assuming $U \subset V$) or there is another data point p such that $p[a_i] = q[a_i]$ ($\forall a_i \in U$) that dominates q on the dimension set $V - U$. Thus, a superset of the union of all subspace skylines is the set of the global skyline points enriched with all points p for which $\exists i \exists q \in SKY_D$ $q[a_i] = p[a_i]$. Consider for example Fig. 6.1 where e and k have the same x-value but k is a subspace skyline point in contrast to point e which is not a skyline point in any subspace. Thus, we can adjust the dominance definition to compute all necessary values co-instantaneously during a skyline calculation. In more details, we define the *extended-skyline (ext-skyline)* based on the *extended domination (ext-domination)* definition [130].

Definition 6.1. Extended Skyline: For any dimension set U, where $U \subseteq D$, p ext-dominates q if on each dimension $d_i \in U$, $p[i] < q[i]$. The ext-skyline (*ext-SKY$_U$*) is set of all points that are not ext-dominated by any other.

In the following, we prove that $ext\text{-}SKY_D$ is sufficient to answer any subspace skyline query correctly.

Lemma 6.1. *Every point that belongs to the skyline of U belongs also to the ext-skyline of U, i.e., $SKY_U \subseteq ext\text{-}SKY_U$.*

Proof: Let $p \in SKY_U$ and $p \notin ext\text{-}SKY_U$. It follows that there is a point q that ext-dominates the point p in U. Based on the definition of the ext-skyline $\forall a_i \in U$: $q[a_i] < p[a_i]$. Therefore, based on the skyline definition we conclude that $p \notin SKY_U$, which leads to a contradiction.

Lemma 6.2. *Every point that belongs to the skyline of a subspace $V \subseteq U$ belongs to the ext-skyline of U, i.e., $SKY_V \subseteq ext\text{-}SKY_U$, $V \subseteq U$.*

Proof: Based on Observation 6.2 we distinguish two cases. If p is a skyline point in U then Lemma 6.1 guarantees that $p \in ext\text{-}SKY_U$. If p is not a skyline point in U there is a skyline point q in U such that $p[a_i] = q[a_i]$, $\forall a_i \in V$. Based on Lemma 6.1 and the definition of ext-skyline we conclude that $q \in ext\text{-}SKY_U$ and $p \in ext\text{-}SKY_U$.

For example, in Fig. 6.1, point m belongs to the ext-skyline, which is not the case with point e, since e is globally dominated by i, which in turn does not have any value equal to the attribute values of e. Notice, that neither e nor m belong to any subspace skyline.

6.2.2 One-dimensional Mapping of Data Points

In order to enable threshold-based pruning of points that cannot possibly belong to the result set, during skyline computation, the multidimensional data is transformed into one-dimensional values by using an appropriate *mapping*. Inspired by [10, 124], each d-dimensional point p is transformed to a one-dimensional value $f(p)$ based on the following formula:

$$f(p) = \min_{i=1}^{d}(p[i]) \tag{6.1}$$

Let $dist_U(p)$ denote the L_∞-distance of point p from the origin, when projected in subspace U, i.e., $dist_U(p) = \max_{i \in U}(p[i])$.

Observation 6.3 *Let p_{sky} be a skyline point in a subspace U. A point p for which the following inequality holds cannot be a skyline point in subspace U.*

$$f(p) > dist_U(p_{sky}) \tag{6.2}$$

Fig. 6.2 depicts a mapping example for a two-dimensional dataset. For sake of simplicity, we assume that the query space U and the data space D are identical. The

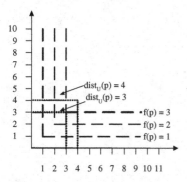

Fig. 6.2 Mapping example

dashed lines correspond to the points with $f(p)$ values of 1, 2 and 3, respectively. Intuitively, by increasing the $f(p)$ values, we are examining the data space in a way that is equivalent to the dashed line shifted from the origin towards the upper-right corner of the data space. In the figure, the dotted lines show the points that cause a threshold value ($dist_U(p_{sky})$) of 3 and 4 respectively. We can see that, for example, if there exists a data point p_{sky} that sets the threshold equal to 3 (that point should lay on the corresponding dotted line), then p_{sky} dominates any point on the dashed line with $f(p)$ equal to 3, independently of the exact location of these points (Observation 6.3).

6.3 Local Data Summaries and Query Processing

6.3.1 Peer and Super-peer Local Data Summaries

In a pre-processing phase, network peers and super-peers locally compute an ext-skyline set of points. This process helps eliminate data items that cannot possibly participate in the result set of a user query from subsequent evaluation. First, each peer P_i computes the local ext-skyline of its dataset S_i and sends it to the associated super-peer. Then, the super-peer gathers the local ext-skylines of its constituent peers and merges them by pruning out those points of a peer P_i that are ext-dominated by points of another peer P_j, resulting in one set that constitutes the ext-skyline on data space D with respect to the portion of the dataset on the super-peer and its associated peers. Based on Lemma 6.1 the local ext-skyline is sufficient for a super-peer to determine if any of its peers P_i can contribute to the results of any skyline query on an arbitrary subspace $U \subseteq D$.

We illustrate the details of peer pre-processing by means of an example. In Fig. 6.3, three peers (P_A, P_B, P_C) assigned to super-peer SP_A and their local datasets are shown. The dimensionality of the dataset is 4. Each peer computes its local ext-skyline in the original space. The points added to the result set due to the ext-

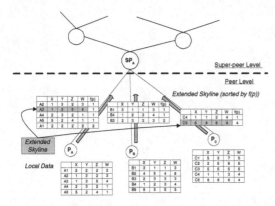

Fig. 6.3 Peer pre-processing example

skyline definition are grey shaded. For instance, four of the five points of P_A are skyline points, while $A3$ is included as an ext-skyline point.

6.3.2 Local Query Processing at Super-peers

Using the locally stored ext-skyline set of points, given a subspace skyline query (denoted by a set of dimensions U), each super-peer is able to answer the query based on its peers' data without actually contacting any of its peers. Still, a substantial amount of computations may be required in order to finally produce the local subspace skyline on U.

6.3.2.1 Indexing of Local Data Summaries

Following the intuition of [124] the ext-skyline points at each super-peer are partitioned in N_C clusters. This can be achieved by using a standard clustering algorithm (like K-Means) or, by employing an application specific clustering method [124]. Each data object is assigned to the nearest cluster based on the distance to the cluster's centroid. The super-peer determines for each cluster C_i the minimum bounding rectangle (MBR_i), which is represented by two reference points l_i and u_i. Point l_i is defined as $l_i[j] = min_{\forall p \in C_i}(p[j])$ and dominates all points in C_i, while point u_i is defined as $u_i[j] = max_{\forall p \in C_i}(p[j])$ and is dominated by all points in C_i. For each data point $p \in C_i$, a one-dimensional mapping is applied according to the distance of p to l_i:

$$f(p,C_i) = \min_{j=1}^{d}(p[j] - l_i[j]) \qquad (6.3)$$

The one-dimensional mapping combined with the clustering information allows us to determine which data points cannot belong to the subspace skyline set for U and can therefore be discarded for this query. Let $dist_U(p, C_i)$ denote the L_∞-distance of point p that belongs to the cluster C_i from the lower corner l_i of the MBR_i based on the dimension set U:

$$dist_U(p, C_i) = \max_{j \in U}(p[j] - l_i[j]) \tag{6.4}$$

Observation 6.4 *Let p_{sky} be a skyline point in subspace U. A point $p \in C_i$ for which the following inequality holds cannot be a skyline point in subspace U.*

$$f(p, C_i) > dist_U(p_{sky}, C_i) \tag{6.5}$$

Based on Observation 6.4, already examined points define a threshold $t(C_i)$ for each cluster C_i, which indicates the region within the cluster that is pruned. Notice that even a point p that does not belong to cluster C_i can dominate points that belong to C_i, as long as p dominates u_i, and thus p can be used to refine the threshold $t(C_i)$.

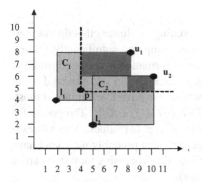

Fig. 6.4 Threshold example

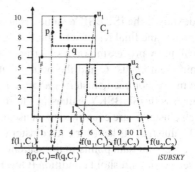

Fig. 6.5 Example of iSUBSKY

Consider for example Fig. 6.4. Point p for which $f(p, C_1) = 1$ sets the threshold $t(C_1)$ based on Equation 6.4 equal to 2. For this threshold value, the pruned area of the cluster C_1 is the dark area within C_1 in the figure. The dominance area of p partially covers cluster C_2, since p dominates u_2 and the threshold value $t(C_2)$ may be refined by p. Hence, the threshold $t(C_2)$ is set to $dist_U(p, C_2)=3$, indicating that point p prunes the dark area in cluster C_2. In this way, we can prune points in cluster C_2 using the updated threshold value $t(C_2)$, before we even start examining the points of cluster C_2.

The one-dimensional values $f(p, C_i)$ refer to the full space (original data space) and do not depend on the queried subspace, therefore they can be computed in a pre-processing phase and stored on disk. Efficient retrieval of these values during query evaluation requires the use of an one-dimensional index, such as the B^+-tree. As will be made clear in what follows, we need a representation of the values where each cluster corresponds to a separate interval in the keys of the B^+-tree. Given a set of N_C clusters and a constant c, each data object p that belongs to a cluster C_i is assigned a one-dimensional iSUBSKY value according to its $f(p,C_i)$ value:

$$iSUBSKY(p) = i * c + f(p,C_i) \qquad (6.6)$$

Using a sufficiently high c value, all objects in the i-th cluster C_i are mapped to the interval $[i * c, i * c + f(u_i, C_i)]$, which is non-overlapping with other cluster intervals. Thus, the use of the iSUBSKY values permits separation of the points of each cluster in the leaves of the B^+-tree as in a clustering index.

Fig. 6.5 depicts an example of the iSUBSKY mapping. Notice that points are mapped into one-dimensional values, whereas clusters correspond to intervals. The actual data points can now be efficiently stored in a B^+-tree using the iSUBSKY values as keys. Additionally, the clusters $C_i = \{l_i, u_i\}$ are kept in a main memory list C. In this way, we can process a subspace skyline query by examining only some specific intervals of the B^+-tree, as described in the following.

6.3.2.2 Subspace Skyline Processing

Indexing of the iSUBSKY values enables early pruning of clusters that do not contribute to the final result set, which reduces the processing cost significantly. During local query processing on a super-peer, the cluster information is accessed from the main memory list C and the points that belong to each cluster are retrieved using the index. For each cluster C_i accessed at query time, we have to examine, at most, points with iSUBSKY values in the interval $[i * c, i * c + f(u_i, C_i)]$. The points for each cluster C_i are examined in an ascending order of $f(p,C_i)$ values. Also a threshold value $t(C_i)$ is kept for each cluster C_i. Notice that most probably we do not have to retrieve all data points in the searched interval, as the processing may stop earlier based on the threshold condition (Observation 6.4).

Algorithm 6.1 describes the pseudocode in detail. For a subspace skyline query $q(U)$, the cluster descriptions C_i are kept in a main memory list C of clusters sorted in ascending order based on the values $\min_{j \in U}(l_i[j])$ (line 3). In each iteration, we examine the next cluster C_m (line 7) and we retrieve the point p that belongs to C_m with the minimum iSUBSKY ($f(p,C_m)$) value (lines 9-10). We repeatedly retrieve the next point based on the iSUBSKY value, until the threshold condition holds (Line 11), since we can safely ignore the remaining points. Each time a new point p is retrieved, we examine if this point is dominated by any point in SKY_U or by any cluster $C_i \in C$ (line 12)[1]. If this is not the case, we remove all points q retrieved so far that are dominated by p (line 13), as well as all clusters C_i that are dominated by p (line 14). Then, p is added to SKY_U as a candidate skyline (line 15). We also refine the current threshold (lines 16-18) of all clusters C_i for which p dominates u_i. The threshold $t(C_i)$ is set as the minimum value of the previous threshold $t(C_i)$ and the new distance $dist_U(p,C_i)$. In the next iteration, if $f(p,C_m) > t(C_m)$ we discard

---

[1] A cluster $C_i(l_i, u_i)$ dominates a point p in U, denoted as $C_i \prec_U p$, if u_i dominates p in U ($u_i \prec_U p$).

Algorithm 6.1 Super-peer local subspace skyline processing

1: **Input:** U: query dimensions
   C: list of clusters $\{C_i\}$
2: **Output:** SKY_U
3: Sort C in ascending order according to the value $\min_{j \in U}(l_i[j])$ for each C_i
4: $SKY_U \leftarrow \emptyset$
5: $t(C_i) \leftarrow$ MAX_INT , $\forall\, C_i \in C$
6: **while** $(C \neq \emptyset)$ **do**
7:      $C_m \leftarrow C$.pop()
8:      $C \leftarrow C - \{C_i\}, C_m \prec_U C_i$
9:      $cursor \leftarrow$ search($[m * c, m * c + f(u_m, C_m)]$)
10:     $p \leftarrow cursor$.pop()
11:     **while** $(f(p, C_m) < t(C_m))$ and $(p \neq$ null$)$ **do**
12:         **if** $(\nexists q \in SKY_U : q \prec_U p)$ and
            $(\nexists C_i \in C : C_i \prec_U p)$ **then**
13:             $SKY_U \leftarrow SKY_U - \{q\}, p \prec_U q$
14:             $C \leftarrow C - \{C_i\}, p \prec_U C_i$
15:             $SKY_U \leftarrow SKY_U \cup \{p\}$
16:             **for** $(\forall C_i \in C : p \prec_U u_i)$ **do**
17:                 $t \leftarrow \max_{j \in U}(p[j] - l_i[j])$
18:                 $t(C_i) \leftarrow \min(t(C_i), t)$
19:             **end for**
20:         **end if**
21:         $p \leftarrow cursor$.pop()
22:     **end while**
23: **end while**
24: **return** SKY_U

C_m without retrieving the remaining points of the cluster from the B^+-tree (line 11). The algorithm returns the subspace skyline set SKY_U (line 24), after examining all clusters that are not dominated by any skyline point retrieved so far.

6.4 Routing Indices

During the pre-processing phase, after computation and clustering of the local ext-skyline set, each super-peer broadcasts the cluster descriptions $C_i(l_i, u_i)$ of its data to the network. This process aims to facilitate the creation of routing indices at super-peer level. As will be explained, these indices will be used for routing queries between super-peers in a manner that aims to decrease the overall query execution cost. Broadcast of the cluster descriptions is considered as a tolerable overhead, since it is a one-time cost (when no updates occur) and it does not saturate the network.

After this exchange, each super-peer holds a list of clusters received through each neighbor. In order to efficiently maintain and process the collected cluster information, we use the iSUBSKY method for indexing the clusters' descriptions. Each super-peer applies a clustering algorithm on all received clusters, resulting in

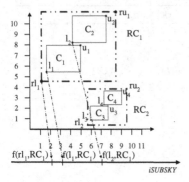

Fig. 6.6 Routing summaries

N_{RC} routing clusters $RC_i(rl_i, ru_i)$, which are represented as MBRs defined by the left lower corner rl_i and the right upper corner ru_i of the MBR. Fig. 6.6 depicts an example where two routing clusters RC_1, RC_2 are created that summarize four clusters C_1-C_4. Merging bounding boxes of two clusters into a new cluster is a well-studied problem from multidimensional indexing methods. Further optimizations of the clustering algorithm are possible but are considered out of the scope of this analysis, therefore in the presented experiments we apply k-Means on the MBRs centers.

The routing clusters RC_i are kept in main memory. For each cluster C_j that belongs to a routing cluster RC_i, we index its lower corner (l_j) using the iSUBSKY value $f(l_j, RC_i)$, and we store for each cluster a triplet $\{f(l_j, RC_i), l_j, u_j\}$. The triples are stored in a B^+-tree using the iSUBSKY values as keys. By accessing the clusters C_j in ascending order of the $f(l_j, RC_i)$ value and setting the threshold based on the maximum distance to the upper corner u_j, it is easy to extend Algorithm 6.1 for subspace skyline computation, where instead of data points, the dominance relationships between MBRs are considered.

6.5 Query Routing and Processing

Using the routing clusters, a super-peer can determine whether forwarding the query to one of its neighbors cannot possibly lead to new results (subspace skyline points). This can lead to significant gains as the corresponding network paths can be safely eliminated from consideration.

Super-peers are only aware of the cluster descriptions of other super-peers. Therefore, it is necessary to process this information at each super-peer and to retrieve the clusters that may contain non-dominated points. In the following, we present an algorithm that takes as input the cluster information, prunes clusters that cannot provide any results and returns only those clusters that may contain non-dominated points. As will be explained, by making use of the compact cluster information, it is feasible to prune parts of other clusters by setting an appropriate threshold value. For that, we assume that a threshold $t(C_i)$ is attached to each cluster C_i. In what follows, we examine the different possible cluster relationships and the conditions that may lead to updating a cluster's threshold. For ease of exposition, we will first describe a simpler version of the algorithm that does not make use of the routing indices.

6.5.1 The SKYPEER Algorithm

The basic algorithm, called SKYPEER, propagates the subspace skyline query to all super-peers and gathers the local subspace skyline results sets using an efficient thresholding scheme. A querying super-peer hands on the query to its neighboring super-peers along the super-peer backbone, which in turn forward the query to their adjacent super-peers. The super-peers execute the query over locally stored extskylines and retrieve local results. These local subspace skyline results are sent back to the initiator super-peer P_{init} (the super-peer that initiated the query) through the query routing path and are merged into a global, final, result set. In the following, we present all relevant steps in detail.

Let t be the threshold value at the end of the local skyline computation at the initiator. The threshold value indicates that there is a local skyline point p that dominates all points with $f(p)$ values larger than t. At the end of the local skyline computation t is set by the point with the minimum threshold value, i.e., the highest pruning capability. Since points are depicted in a common data space in all super-peers, the local skyline point p dominates points in other super-peers as well, i.e., those with $f(p)$ larger than t. Therefore, t is attached to the query by the initiator and is propagated to the rest of the super-peer network. The threshold value is used as an initial threshold for the local subspace skyline computation at recipient super-peers to further reduce the computation and communication cost.

An alternative threshold propagation strategy is to refine (if possible) the threshold at each super-peer that is processing the query, and then propagate the refined value to adjacent super-peers, instead of forwarding P_{init}'s fixed threshold value. Intuitively, by progressively lowering the threshold value, the pruning capability of the query increases at each forwarding step. However, this approach requires that the query is propagated only after the super-peer has finished the local skyline computation.

During query processing, the local subspace skyline sets have to be merged into one overall result set. We explore two different strategies for the merging phase. The simplest strategy is that the initiator super-peer P_{init} collects the local subspace skyline result sets of all super-peers and merges the local skylines into one global result set. Even though the use of the threshold reduces the amount of data transmitted to the query initiator, there is still a chance that a local result may contain points that do not belong to the overall skyline. Therefore, an alternative strategy for the merging phase, is to merge progressively the local skyline sets during query evaluation.

| Fixed Threshold Fixed Merging (FTFM) |
| Fixed Threshold Progressive Merging (FTPM) |
| Refined Threshold Fixed Merging (RTFM) |
| Refined Threshold Progressive Merging (RTPM) |

Table 6.1 Variants of the basic SKYPEER

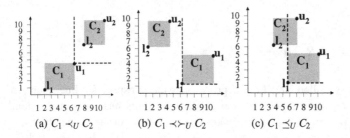

(a) $C_1 \prec_U C_2$ (b) $C_1 \prec\succ_U C_2$ (c) $C_1 \preceq_U C_2$

Fig. 6.7 Clusters domination examples

Instead of forwarding all results back to P_{init}, each super-peer merges the results of its neighbors, and forwards the merged result back to P_{init}. The expected benefit of this alternative merging strategy is twofold. First, the amount of transferred data is reduced because of the progressive merging of the local result sets. Moreover, a time-consuming centralized merging process at the query initiator is avoided, as this task is essentially spread amongst all super-peers.

In Table 6.1 we summarize the four variants of the SKYPEER algorithm based on (i) whether a fixed threshold induced by the initiator is used versus progressively refining the threshold at each super-peer and (ii) whether a fixed merging policy (at the initiator) is chosen over progressive merging of the results.

6.5.2 Cluster Dominance Relationships

In order to facilitate intelligent propagation of a query in the super-peer network, using the routing indices, we need to derive a process for refining the thresholds that will be used to prune clusters (and thus super-peers) that cannot contribute to the result set. Moreover, refinement of the thresholds allows pruning of local clusters from further consideration, reducing this way the local processing cost in a super-peer.

Based on the employed cluster representations, we can straightforwardly derive dominance relationships between clusters. Given two clusters $C_1(l_1, u_1)$ and $C_2(l_2, u_2)$ we define the following dominance relationships (also depicted in Fig. 6.7).

Definition 6.2. Cluster C_1 dominates C_2 in U, denoted as $C_1 \prec_U C_2$, if u_1 dominates l_2.

Algorithm 6.2 Threshold update $(C_n, \{C_i\})$

1: **Input:** C_n: cluster
$\quad$ C: list of clusters $\{C_i\}$
2: **Output:** updated thresholds
3: **for** $(\forall C_i \in SKY_U : C_n \preceq_U C_i)$ **do**
4: $\quad$ $t \leftarrow \max_{j \in U}(u_n[j] - l_i[j])$
5: $\quad$ $t_i \leftarrow \min(t_i, t)$
6: **end for**
7: **for** $(\forall C_i \in SKY_U : C_i \preceq_U C_n)$ **do**
8: $\quad$ $t \leftarrow \max_{j \in U}(u_i[j] - l_n[j])$
9: $\quad$ $t_n \leftarrow \min(t_n, t)$
10: **end for**

If C_1 dominates C_2, this means that there exists at least one point in C_1 that dominates all possible points in C_2. Cluster C_2 cannot contribute to the subspace skyline result and can be safely pruned during query processing.

Definition 6.3. Clusters C_1 and C_2 are incomparable in U, denoted as $C_1 \prec\succ_U C_2$, if l_1 does not dominate u_2 and l_2 does not dominate u_1.

In this case, no point of C_1 can be dominated nor dominate any point of C_2, thus both clusters have to be examined during query processing.

Definition 6.4. Cluster C_1 partially dominates C_2 in U, denoted as $C_1 \preceq_U C_2$, if l_1 dominates u_2 but u_1 does not dominate l_2.

Therefore if C_1 partially dominates C_2 some points of C_1 may dominate some points of C_2. In this case, we have to examine both clusters in order to retrieve the entire subspace skyline set. However, we can set the threshold of C_2 corresponding to the intersection of the pruning area of C_1 with C_2.

Observation 6.5 *Given two clusters C_1, C_2 where $C_1 \preceq_U C_2$, if $u_1 \prec_U u_2$ then the threshold $t(C_2)$ can be set as $dist_U(u_1, C_2)$.*

In the case where C_1 partially dominates C_2 in U, we can adjust the threshold of C_2, if u_1 dominates u_2. Algorithm 6.2 describes the procedure of updating the thresholds. Each time a threshold is updated, we keep the minimum value of the current threshold and the previously defined threshold by another dominance relationship.

The dominance relationships between clusters are used by the routing algorithm for either pruning entire clusters or restricting the number of points in a cluster that should be accessed, as will be shown shortly.

6.5.3 The SKYPEER+ Algorithm

We now present the details of the SKYPEER+ algorithm, which facilitates intelligent routing of the query in the super-peer network and effective pruning of local clusters, based on the information contained in the routing indices.

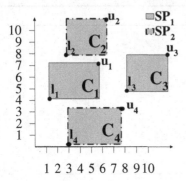

Fig. 6.8 Clusters of super-peers SP_1 and SP_2

During query processing, each super-peer that receives a subspace skyline query 1) uses the query routing algorithm (Algorithm 6.3) to identify local and non-local candidate clusters, 2) propagates the query to the neighboring super-peers responsible for the non-local clusters, and 3) processes the query locally for the local candidate clusters using Algorithm 6.1. In the following we describe these steps in detail.

First, let us consider a simple example depicted in Fig. 6.8. Both super-peers SP_1 and SP_2 eventually store the cluster descriptions of all four clusters. For better readability, we depict only the clusters and not the routing clusters. Let us assume that SP_1 and SP_2 store incomparable clusters (namely C_1 and C_4), but also some dominated or partially dominated clusters by other super-peers. If SP_1 receives a two-dimensional skyline query, SP_1 concludes based on the routing clusters that the query has to be propagated to SP_2, because of the clusters C_2 and C_4. Furthermore, SP_1 starts a local skyline computation only for cluster C_1, while cluster C_3 is discarded since it is dominated by cluster C_4 of SP_2. When super-peer SP_2 receives the query, it first examines the routing information, and sets a threshold for cluster C_2. The threshold is set based on cluster C_1 of SP_1 and corresponds to the pruning area of point u_1. Thereafter, a skyline computation on clusters C_2 and C_4 is initiated. Notice that the threshold for cluster C_2 cannot be refined based on cluster C_4, even though C_2 is partially dominated by C_4, since point u_4 does not dominate point u_2. In this way, each super-peer selectively processes only those clusters, that can affect the result set, using appropriate threshold values.

Algorithm 6.3 describes the proposed routing algorithm. As in iSUBSKY method, the routing clusters $RC_i = [rl_i, ru_i]$ are examined in ascending order of the min $_{j \in U}(rl_i[j])$ values (lines 3-7). During processing, clusters (line 13) and routing clusters (line 14) that are dominated by the currently examined cluster C_n are discarded. For clusters and routing clusters that are partially dominated by or partially dominate C_n, we compute new threshold values (line 15 and 17, respectively). At the end of Algorithm 6.3, C_{SKY} contains all non-dominated clusters that should be processed.

Following execution of the algorithm the clusters in C_{SKY} are examined. If cluster $C_i \in C_{SKY}$ is a local cluster, we add C_i to the list C ($C = C \cup C_i$) with clusters that should be processed locally and set the initial threshold of C_i based on Algo-

---

Algorithm 6.3 Super-peer query routing

1: **Input:** U: query dimensions
   RC: list of routing clusters RC_i
2: **Output:** C_{SKY}: list of non-dominated clusters
3: Sort RC in ascending order according to the value $\min_{j \in U}(rl_i[j])$ for each RC_i
4: $C_{SKY} \leftarrow \emptyset$
5: $t(RC_i) \leftarrow$ MAX_INT, $\forall RC_i \in RC$
6: **while** $(RC \neq \emptyset)$ **do**
7:     $RC_m \leftarrow RC.\text{pop}()$
8:     $RC \leftarrow RC - \{RC_i\}, RC_m \prec_U RC_i$
9:     $cursor \leftarrow \text{search}([m*c, m*c + f(ru_m, RC_m)])$
10:    $C_n \leftarrow cursor.\text{pop}()$
11:    **while** $(f(l_n, RC_m) < t(RC_m))$ and $(C_n \neq \text{null})$ **do**
12:       **if** $(\nexists C_i \in C_{SKY} : C_i \prec_U C_n)$ and
          $(\nexists RC_i \in RC : RC_i \prec_U C_n)$ **then**
13:          $C_{SKY} \leftarrow C_{SKY} - \{C_i\}, C_n \prec_U C_i$
14:          $RC \leftarrow RC - \{RC_i\}, C_n \prec_U RC_i$
15:          updateThreshold(C_n, C_{SKY})
16:          $C_{SKY} \leftarrow C_{SKY} \cup C_n$
17:          updateThreshold(C_n, RC)
18:       **end if**
19:       $C_n \leftarrow cursor.\text{pop}()$
20:    **end while**
21: **end while**
22: **return** C_{SKY}

---

rithm 6.2. Otherwise, the subspace skyline query has to be posed to the corresponding super-peer $SP(C_i)$, therefore we add $SP(C_i)$ to the list SP ($SP = SP \cup SP(C_i)$) of super-peers that should be contacted. Then the super-peer propagates the query to all $SP_i \in SP$ and processes the query locally (for the clusters in list C), by using Algorithm 6.1.

Notice that in contrast to SKYPEER, we do not need to propagate any threshold among super-peers during query processing. Instead the threshold is refined at each super-peer based on the routing information, i.e., the cluster descriptions. Similar to SKYPEER, the intermediate results have to be merged as described in Section 6.5.1, which leads to two different variants of SKYPEER+. In the case of fixed merging at P_{init} we have the RFM (Routing-based Fixed Merging) variant, while progressive merging leads to RPM (Routing-based Progressive Merging).

6.6 Other Existing Approaches

Finding skyline points was first investigated in computational geometry as maximal vector problem [89, 14]. Recently skyline computation has received considerable attention in the database research community. Börzsönyi et al. [21] first investigate the skyline computation problem in the context of databases. In the following, we present a short overview of existing approaches for skyline computation assuming first a centralized and then a distributed setting.

6.6.1 Centralized Algorithms

The first family of skyline algorithms is based on *sequential scanning* of the dataset. In [21], two algorithms BNL (Block-Nested-Loop) and D&C (Divide-and-Conquer) are proposed. The BNL algorithm uses a block nested loop to compare each point of the database with every other point and reports it as a result only if it is not dominated by any other point. The D&C algorithm divides the data space into several regions repetitively, calculates the skyline in each region, and produces the final skyline from the points in the regional skylines. SFS (Sort-First-Algorithm) [36], is based on the same principle as BNL, but improves performance by first sorting the data according to a monotone function. Godfrey et al. then proposed a skyline algorithm LESS [62] based on SFS, which achieves a better average performance. In [10] the idea of limiting the amount of data to be read by exploiting the value of a monotone function was studied. A more efficient algorithm termed Lattice Skyline (LS), which is applicable in the case of low cardinality domains is proposed in [98].

Several *index-based* techniques are proposed in the relevant research literature. Tan et al. [122] discuss algorithms for progressive skyline computation. An efficient algorithm that uses nearest neighbor search in a dataset indexed by an R-tree, was proposed in [86]. Later, Papadias et al. [102, 103] propose a branch and bound algorithm (BBS) to progressively output skyline points on a dataset indexed by an R-tree, with guaranteed minimum I/O cost. Recently the ZBTree [90] index structure, which is based on Z-order curve, is shown to outperform BBS.

Existing approaches have not only studied efficient skyline computation, but also proposed variations of the traditional skyline operator. Motivated by the fact that different users may issue queries regarding different subsets of the available attributes depending on their interests, recent papers focus on algorithms to support *subspace skyline* retrieval. In [124, 125], the SUBSKY algorithm is presented, which transforms the multidimensional data to one-dimensional values, and then indexes the dataset with a B-tree. In [47] the problem of supporting constrained subspace skylines was posed. Pei et al. [105] and Yuan et al. [145] first defined the union of the skyline sets in all possible subspaces, as a new operator. Pei et al. [105] discussed subspace skylines primarily from the view of query semantics. They presented the skyline membership query, namely why and in which subspaces an object belongs to the subspace skyline, by using the notion of skyline group. The authors in [145] present a pre-processing approach, called SKYCUBE, which is defined as the union of all skyline points of all possible non-empty subspaces. For this purpose, they explore sharing strategies for answering multiple subspace skyline queries by identifying computational dependencies among subspace skyline sets. Recently, Xia and Zhang [139] address the issue of supporting updates in SKYCUBE by introducing the compressed SKYCUBE. The efficient computation of the compressed SKY-

CUBE was also studied in [104]. In [83] an approach that relies on materialization of dominance relationships in subspaces is presented. The main idea is to pre-compute dominance relationships between pairs of points in subspaces and organize them to build the maximal space index (MS-index).

The *cardinality* of the skyline result set was studied in [61, 29]. It has been shown that, for a random dataset, the expected number of skyline points is $\Theta(\ln^{d-1} n/(d-1)!)$. This result verifies the observation that the skyline cardinality increases significantly with the dataset dimensionality. Therefore, recent papers focus on *variations of the skyline definition* or *ranking of skyline points*, in order to determine a small set of the most *representative* skyline points.

Chan et al. [26] propose the k-dominant skyline query to restrict the skyline cardinality. The authors relax the idea of dominance to k-dominance, in order to increase the probability of one point dominating another point. In [91] the authors study the problem of selecting k skyline points so that the number of points, which are dominated by at least one of these k skyline points is maximized. Koltun and Papadimitriou [85] first confront the issue of approximate domination from a theoretical perspective. Then, Xia et al. [140] present a similar idea, the ε-skyline, which essentially relaxes the dominance relationship by adding a constant value ε to all dimensions.

Ranking of skyline points was discussed in [27]. The authors introduce a new metric called skyline frequency, to compare and rank the interestingness of data points based on how often they are returned in the skyline, when different subspaces are considered. To avoid the $2^d - 1$ subspace skyline computations that are required, they propose an approximate algorithm for estimating the skyline frequency. Moreover, several works discuss ranking of skyline points relying on user-defined functions or preferences on some dimensions. In [79], the authors present the Telescope algorithm that ranks the skyline points by user-specified preferences on the available dimensions. In [9], a ranking approach based on user -defined regions that dominate all other regions is proposed. In [135], a framework for ranking skyline points in the absence of a user-defined preference function is proposed, that relies on the skyline graph that is defined based on the dominance relationships between the skyline points for different subspaces.

6.6.2 Distributed Algorithms

There has been a growing interest in distributed skyline computation. In [5, 93], skyline processing over distributed web information systems was studied and three different algorithms have been proposed, namely the basic distributed skyline (BDS) algorithm [5], the improved distributed skyline (IDS) algorithm [5], and the progressive distributed skyline (PDS) algorithm [93]. Unfortunately, their assumptions are hardly applicable to large-scale P2P systems. Furthermore, while they assume vertical partitioning of the dataset across the Web sources, data in P2P systems is, typically, horizontally partitioned across the peers.

Hose et al. [73] propose an approach for skyline processing in unstructured P2P networks that uses routing indexes to identify relevant peers. The proposed multi-dimensional routing indices are based on the Q-tree, which is a fusion of R-trees and histograms. The proposed techniques provide probabilistic guarantees for the result's correctness. Huang *et al.* [76] assume a setting with mobile devices communicating via an ad-hoc network (MANETs), and study skyline queries that involve spatial constraints. The authors present techniques that aim to reduce both the communication cost and the execution time on each single device. In [134], bandwidth-constrained skyline queries in mobile environments are studied.

There are also several approaches for P2P skyline computation that apply space partitioning techniques. Wang et al. [136] use the Z-curve method to map the multidimensional data space to one-dimensional values, that are assigned to peers connected in a tree overlay like BATON [81]. Later, the authors present SkyFrame [137] as an extension of their work. In this approach, due to the space partitioning scheme, a load balancing problem may arise. In particular, a small number of peers (those that are allocated space near the origin of the axes) eventually have to process almost every query. In order to improve the performance the authors propose extensions such as smaller space allocation to peers responsible for these regions or data replication techniques. Chen et al. [35, 43] propose the iSky algorithm, which employs an alternative transformation, namely iMinMax, in order to use the BATON overlay.

In addition, a few approaches have been proposed that assume data partitioning among servers without the restriction of an existing overlay network, i.e., the query originator can directly communicate with all servers. Cui et al. [44, 34] proposed the PaDSkyline algorithm, where the data stored at each server are summarized by MBRs. SkyPlan was proposed in [109] for improving the performance of PaDSkyline. SkyPlan addresses the problem of generating execution plans, has a direct impact on the performance of skyline query processing. Assuming the same architecture, in [108], an approach called AGiDS is proposed that uses a grid-based data summary of the data stored locally at each server. In [148], a feedback-based distributed skyline (FDS) algorithm is proposed, which aims to minimize the bandwidth consumption at the expense of several roundtrips.

In [138], Wu et al. first address the problem of parallelizing skyline queries over a shared-nothing architecture. The proposed algorithm named DSL, relies on space partitioning techniques. The author propose two mechanisms, recursive region partitioning and dynamic region encoding. Their techniques enforce the skyline partial order, so that the system pipelines participating machines during query execution and minimizes inter-machine communication.

Parallel skyline computation is studied in [40]. The algorithm first partitions the dataset in a random way to the participating machines, in order to ensure that the structure of each partition is similar to the original dataset. Then each machine processes the skyline over its local data using an R-tree as indexing structure. A different approach regarding parallel skyline computation is presented in [59]. In contrast to the previous approaches, the authors assume a multi-disk architecture with one processor and they make use of the parallel R-tree. The main focus of this paper is to access more entries from several disks simultaneously, in order to improve the pruning of non-qualifying points. The authors focus on the efficient distribution of the nodes of the parallel R-tree.

6.7 Summary

In this chapter, we studied the problem of efficient skyline computation in a P2P setting. A main objective during distributed skyline processing, is to only contact peers that store relevant data. Peers that store only dominated data points should not be queried at all, while data points that are dominated by points stored at another peer should not be transferred. These decisions are hard to make when no global knowledge of the data distribution is available. We presented in details a distributed framework for subspace skyline processing, called SKYPEER+, that achieves these goals by employing a thresholding scheme that detects (locally and globally) dominated data points and eliminates them from further consideration. Furthermore, we analyzed an appropriate routing mechanism for subspace skyline queries over a super-peer network that relies on local data summarization. Thus, the skyline set is retrieved by contacting only those super-peers that may contribute to the overall subspace skyline result set, reducing both the computational cost and the amount of transferred data.

Chapter 7
Top-k queries

Abstract Recently there has been an increased interest in database management systems to incorporate and support more flexible query operators, such as top-k, that produce results of specified cardinality, thus avoiding huge and overwhelming result sets. Top-k queries retrieve the objects that best match the user requirements by employing user-specified scoring functions that result in an ordered set of objects containing the best k objects only [30, 75]. In this chapter, efficient processing of top-k queries in peer-to-peer systems is studied. To this end, the applicability of the skyline operator is investigated for efficiently answering top-k queries for a wide class of scoring functions, indicating user-specified preferences, in large P2P networks.

7.1 Overview

A number of applications can significantly benefit from support for top-k query processing, for example multimedia retrieval (including images) [31, 63], digital libraries [94, 95], web search [97], and e-commerce [96]. Consider for example online booking systems, e.g, travel and accommodation, where the user is only interested in the best offers (air-tickets, hotels) according to a set of dynamic, user-specified criteria. Due to applications and systems such as sensor networks, data streams, and peer-to-peer (P2P) systems, data generation and storage is becoming increasingly distributed. Thus an emerging challenge is to support top-k query processing over a highly distributed network of collaborative computers (i.e., servers or peers).

The main focus of this chapter is on top-k query processing in P2P systems. Each user may define his/her own arbitrary preferences for each query, therefore the queries are not necessarily re-occurring. The challenge is to provide efficient algorithms for processing top-k queries, i.e., queries that return only the exact best k results to the user. SPEERTO is a framework that supports top-k query processing over horizontally partitioned data stored on peers organized in a super-peer network. Users are allowed to specify a monotone function for each query that aggregates a certain number of the objects characteristics into a single *score* that defines a total ordering, and therefore enables the retrieval of top-k results. SPEERTO supports a large class of scoring functions and uses the skyline set [21] for answering top-k queries. For a maximum value of K, denoting an upper bound on the number of results requested by any top-k query ($k \leq K$), each peer computes its K-skyband [103] as a pre-processing step. Each super-peer maintains and aggregates the K-skyband sets of its peers to answer any incoming top-k query. By exchanging skyline sets (a skyline is a subset of the K-skyband set) at super-peer level, SPEERTO always provides the exact and complete result set in a progressive way, while queries are deliberately routed to those super-peers that actually contribute to the top-k result.

The rest of this chapter is organized as follows: In Section 7.2, the local data summaries and query processing are described. In Section 7.3, the construction of the skyline-based routing mechanism for top-k query processing over a super-peer architecture is presented. Thereafter, in Section 7.4 the threshold-based top-k algorithm is presented. Section 7.5 reviews the related work, and finally Section 7.6 provides a brief summary of the contents of this chapter.

7.2 Local Data Summaries and Query Processing

The result of top-k queries for any increasingly monotone function can be answered using the K-skyband (where $k \leq K$). The K-skyband is a set of points, such that there exists no other point that can belong to the result of any top-k query for any increasingly monotone function. Therefore, the K-skyband can be used as data summary of a peer's data, in the case of top-k queries with $k \leq K$.

When a peer joins the P2P network, the peer computes the K-skyband of its local data. Each super-peer gathers the K-skyband sets from its peers and merges the individual K-skyband sets by discarding points that are dominated by more than $K - 1$ points. In this way a super-peer maintains the aggregated K-skyband set of all data stored at its peers, and each super-peer is capable to answer any incoming top-k query over its peers' data.

It should be stressed that even though the skyline operator and the K-skyband are CPU-intensive [29] and therefore more costly than a top-k query, they are only computed as a pre-processing step, i.e., their construction is a one-time cost, and then any top-k query with arbitrary k ($k \leq K$) and scoring function can be processed. Thus, during local query processing a super-peer that is queried executes a top-k query on the locally stored K-skyband points.

7.3 Routing Summaries

Given the K-skyband at each super-peer, there exist two naive solutions to process top-k queries over the super-peer network. In the first, each super-peer broadcasts its

K-skyband to all other super-peers, then each super-peer has enough data to answer any top-k ($k \leq K$) query locally. The advantage is that the query can be processed (at any super-peer) without contacting remote super-peers. However, this approach is not feasible in a highly distributed environment, because of the size of the skyband and the cost of distributing it to all the super-peers and keeping it updated.

The second naive approach is to flood each query to all super-peers to find the correct top-k result. The advantage of this approach is that the cost of distributing the skybands is avoided. However, flooding is costly, and although appropriate for distributing metadata needed for creating routing indices, it is too costly to employ for each individual query.

A more efficient approach than the naive approaches outlined above, which combines the advantages of the aforementioned approaches while alleviating the disadvantages, is for each super-peer to broadcast only some summary information of its K-skyband, namely the skyline set. Intuitively, the skyline is the border of the K-skyband with respect to the axes. Notice that the cardinality of the skyline is significantly smaller than the cardinality of the K-skyband. The skyline set is sufficient for any super-peer to route any top-k query only to those super-peers that can contribute to the final result. Each super-peer SP_i assembles N_{sp} sets of skyline points SKY_i, ($1 \leq i \leq N_{sp}$). These points are called *routing objects*.

The cardinality of the skyline set [29] depends on the data distribution and the data dimensionality and influences the performance of distributed top-k processing. If the skyline set is of high cardinality, it may cause high construction and maintenance cost and storage requirements at super-peers. SPEERTO can be easily extended to support data summaries of fixed size. In [133], the problem of finding an approximation of the skyline of fixed size to distribute to all super-peers as a routing mechanism is studied. The goal is to approximate and replace the skyline set with a set of points of fixed size, while the routing ability of the approximation is influenced as little as possible. Reducing the number of the routing objects may lead to more contacted super-peers and more transferred data, however it also reduces the construction and maintenance costs, which is also important.

An advantage of using the skyline as summary information is that data updates change the skyline rarely, and small changes in the skyline do not significantly change the accuracy of the top-k query processing. Therefore, it is not necessary to continuously maintain the skyline updated at remote super-peers, and periodic updates suffice. Obviously, high update rates can lead to time intervals where the results may not be accurate temporarily. The maintenance approach of remote skylines is based on broadcasting the skyline updates, when either the skyline has significantly changed or the validity time has expired. While skylines are used to select super-peers during top-k processing, the K-skyband on a super-peer is used to generate the actual results, and has to be more frequently updated. However, this cost is still less significant because a peer is relatively close to its super-peer in terms of network distance.

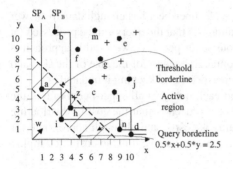

Fig. 7.1 Illustration of query and threshold borderline and active region

7.4 Query Routing and Processing

In the following, we describe a threshold-based top-k algorithm assuming that every super-peer SP_i ($1 \leq i \leq N_{sp}$) stores the routing objects SKY_j of all other super-peers ($1 \leq j \leq N_{sp}$) and the merged K-skyband set ($KSKY_i$) of its associated peers only.

Let us first consider a two-dimensional space as a showcase scenario, as illustrated in Fig. 7.1. In the figure, the depicted routing and data objects are stored at super-peer SP_A. In more details, the skyline sets of two super-peers SP_A and SP_B are shown along with the routing objects from the other super-peers. For clarity reasons, we label only the routing objects of the two super-peers SP_A and SP_B that are actually involved in the top-k query of the example. Routing objects are depicted as black circles, while data objects are marked with crosses. The data objects are the points that belong to the $KSKY_A$, i.e., the aggregated K-skyband set of super-peer SP_A. Therefore, Fig. 7.1 depicts the information that is available to super-peer SP_A when query processing starts. As described in more detail below, during query processing more data objects are transferred to the querying super-peer SP_A through the neighboring super-peers.

Given an arbitrary weighting vector that defines the user's top-k query (i.e., the user's preferences), a perpendicular line to this vector can be uniquely defined. This line is also called *query borderline* in the following. In a two-dimensional space, progressive processing a top-k query can be visualized by sweeping the query borderline, with specific slope defined by the query weights, through space from the axes towards the data. The first data point that the line meets is the top-1, the second the top-2, etc. until it finds top-k data points. Actually, each time the line meets a data point, this point can be immediately returned to the user, as it is really the next top object of the query (progressive property of the algorithm). Note that some of the points that the query borderline meets are routing objects. In this case, each routing object met must be replaced by some data points of the super-peer to which the routing object belongs to. At each step the query borderline is an indication of how far the data space is examined and it guarantees that there does not exist any other

---

Algorithm 7.1 Query processing on super-peer SP_Q

1: **Input:** Query $q_k(f)$
2: $list = \emptyset$
3: $list = SP_Q.\text{query}_{\cup SKY_i}(q_k(f))$
4: $threshold = f(list[k])$
5: $c = 0$
6: **while** $(c < k)$ **do**
7:     $next\_obj = list.pop()$
8:     **if** $next\_obj$ is a *routing object* **then**
9:         $SP = next\_obj.super\text{-}peer()$
10:         $temp = SP.\text{query}(q_{k-c}(f), threshold)$
11:         $list.\text{removeRoutObj}(SP)$
12:         $list.\text{add}(temp)$
13:     **else**
14:         return $next\_object$ to the user
15:         $c = c + 1$
16:     **end if**
17:     $threshold = f(list[k-c])$
18: **end while**

---

point in the examined space that has not been retrieved yet. This guarantees that there exists no data point that has a better scoring value than the retrieved points.

A threshold value is defined as the score of the k-th routing or data object encountered so far. In the two-dimensional space, this defines a *threshold borderline* that gradually sweeps the space towards the axes origin. The region defined between the query and the threshold borderline is called *active region* and it contains at least $(k - c)$ objects, where c is the number of data points that is already returned to the user. In each step, the active region contains all objects that may appear in the final result set. Notice that the querying super-peer is not aware of all data points that fall in the active region, therefore if a routing object is retrieved, the query must be broadcast to the corresponding super-peer.

In order to take a concrete example based on the information depicted in Fig. 7.1, consider a linear top-4 query with weights $w = (0.5, 0.5)$. Let us further assume that the query is posed at super-peer SP_A. The top-4 objects (i, a, h and z) based on the data and routing objects stored on SP_A are retrieved, and the score of the 4-th object (z), defines the threshold borderline. This guarantees that the results of this top-k query are found in the active region. Notice that some data points of SP_B may fall in the active region and therefore point z may not belong to the top-4 result set. First, the routing object i is examined and since it belongs to SP_A the data object i is retrieved and returned to the user. In the next step, point a is retrieved and returned as the top-2 point. Afterwards, the routing object h is retrieved that belongs to SP_B. Therefore, super-peer SP_B is queried and assuming that no other data point of SP_B falls in the active region, points h and z are returned to the user.

Algorithm 7.1 describes how P2P top-k query processing is performed. The routing and data objects retrieved thus far are kept in a sorted list based on the scoring value. This list is initialized by the querying super-peer (SP_Q) with the top-k objects of the skyline results $\cup SKY_i$. The scoring value of the k-th object is used as a threshold, since any other object with higher score cannot belong to the final result set. In each iteration the top object of the list is examined. Then, a top-k query is broadcast to the super-peer (SP) responsible for this object. After SP's data objects are retrieved by SP_Q, all routing objects of SP are removed from the sorted list before inserting its data objects, since they are no longer necessary to maintain. Afterwards, the threshold is updated with the scoring value of the k-th object in the list. In each subsequent iteration, if a data object is retrieved, it is returned to the user as the top-1, top-2, etc. result. Otherwise, if a routing object is retrieved, a top-$(k-c)$ query is send to the corresponding super-peer along with the current threshold value (c denotes the number of results returned thus far to the user). The super-peer sends back $k-c$ objects, or less if there are not $k-c$ objects with value below the threshold. The algorithm terminates when k data objects have been retrieved from the sorted list.

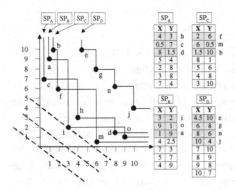

Fig. 7.2 Example of top-k algorithm

Example: Consider a small super-peer network consisting of four super-peers $SP_A, \ldots, SP_D$, and a querying super-peer $SP_Q = SP_A$ that has assembled the skylines (routing objects) of the other super-peers, as depicted in Fig. 7.2. Let us assume that SP_A needs to answer a top-3 query with a linear aggregate function that assigns equal weights to both dimensions. On the right side of the figure, the skyband information maintained on each super-peer is depicted in tables. The grey-shadowed objects are the skyline objects that are broadcast to other super-peers and they are also depicted on the left part of the figure graphically. According to Algorithm 7.1, the sorted list is initialized with routing objects: $i(3,2)$, $m(6,0.5)$ and $h(4,3)$ and threshold is set to 3.5. The first object that is processed is object $i(3,2)$ that belongs to super-peer SP_B. Thus SP_A sends a top-3 query to SP_B, and retrieves its local (at SP_B) top-3 results. These data objects are $i(3,2)$, $(4,2.5)$ and then for the third ranked object there are actually three objects with the same aggregate score $o(9,1)$, $a(1,9)$ and $(7,3)$. These three data objects are not returned to SP_A since they are discarded by the threshold value. So, only two points are returned to SP_A by SP_B and they are merged with the objects already existing in the list. The threshold value is set to 3.25, as the new k-th object (4,2.5) has a lower score value than the old threshold value. Then $i(3,2)$ is returned to the user as top-1. The list contains two objects: $m(6,0.5)$, (4,2.5). Next m is processed and since it belongs to SP_C, SP_A sends a message to SP_C requesting its top-2 objects. SP_C returns $m(6,0.5)$, while $f(2,6)$ is pruned by the threshold. Thereafter, m is returned to the user and the list contains only one object: (4,2.5). This object is processed next, and since it is a data object, it is returned to the user immediately as the top-3. Finally, the algorithm terminates.

Correctness and Optimality. The usage of the threshold guarantees that SPEERTO progressively returns accurate and exact answers for any top-k ($k \leq K$) query. Moreover, it reduces communication costs by preventing unnecessary data objects from being transferred in the network during query processing. SPEERTO also avoids querying super-peers that do not contribute to the result set. Assuming that the query is answered using a snapshot of the P2P network, i.e., static network and contents, SPEERTO minimizes the number of contacted super-peers and the amount of transferred data during query processing, while always retrieving the correct result set.

7.5 Other Existing Approaches

Rank-aware query processing is a cornerstone process for many applications, including ranked document retrieval from digital libraries and the Web. Top-k queries aim at identifying the k most relevant data items with regards to the user's preferences. In [24], the authors extend SQL with explicit support for limiting the cardinality of a query result to a user-specified number of tuples. Ilyas et al. [78] provide a survey of existing top-k query processing techniques and classify the different methods into a taxonomy.

7.5.1 Centralized Algorithms

Several papers have dealt with the issue of top-k query processing in centralized database management systems. Onion [28] supports top-k queries efficiently by computing and storing in a pre-processing phase the convex hulls of data points in layers, with outer layers geometrically enclosing the inner ones. A top-k query for any linear preference function can be processed, based on the observation that the point with the highest score can be found within the convex hull of the dataset. Therefore, Onion evaluates a linear top-k query by processing the layers inwards, starting from the outmost hull.

Another pre-processing based technique for top-k queries is Prefer [75]. Prefer uses materialized views for top-k query processing. The proposed method materializes the top-k result sets as views, according to some arbitrary scoring functions. During query processing, for any given preference function, Prefer selects the materialized view corresponding to the function that is most similar to the querying scoring function. Then, the query can be answered by examining a subset of the data elements in this view. The main challenge is to answer efficiently top-k queries using a reasonable number of materialized views. Prefer works for non-linear scoring functions, provided that a different set of views is maintained for each function type. Onion and Prefer are mostly appropriate for static data, due to the high cost of pre-processing.

Efficient maintenance of materialized views for top-k queries is discussed in [142]. The authors propose algorithms that reduce the storage and maintenance cost of materialized top-k views in the presence of deletions and updates. Based on the the observed workload, the proposed method uses a system parameter k_{max} and during the pre-processing phase, a top-k_{max} query is processed and the data elements are materialized as a view. Incoming tuples (updates or insertions) with score larger than the score of the k-th tuple are inserted into the view. On the other hand, deleted tuples may reduce the number of entries, and then, a top-k_{max} query is issued again on the database to update the view.

The authors in [22] study the advantages and limitations of processing a top-k query by utilizing multidimensional histograms and translating queries into a single range query that a traditional relational database management system can process efficiently. In [33], a method is proposed that transforms a top-k query into an approximate range query by using a sampling-based approach, along with various query mapping strategies, to determine a range query that yields high recall with low access cost. In both previous approaches, if the range does not return at least k results, a new range query has to be posed.

Recently, reverse top-k queries [128, 129] have been introduced, as a query that identifies the scoring functions (i.e., weighting vectors) that make a point a top-k result. Extensions of reverse top-k queries have also been proposed for mobile environments [131] and for identifying influential products [132].

7.5.2 Distributed Algorithms

Several papers focus on computing the top-k queries on vertically distributed data over multiple sources, where each source provides a ranking over some attributes. The query is submitted to d sources and each source returns a list of objects sorted in descending order of their partial scores with respect to the corresponding attributes. The problem is to compute the top-k results in terms of their overall score by combining the d sorted lists. The existing algorithms can be distinguished in two categories, based on whether they assume only *sorted accesses* or also *random accesses*, i.e., the score of any random data element in the list can be immediately returned. Fagin et al. [54] introduce two algorithms, namely TA and NRA algorithms. The TA algorithm is optimal in terms of random accesses, for repositories that support random access, while the NRA algorithm assumes that only sorted access is available. Variations of the methods have been proposed that try to improve some limitations of the aforementioned methods and examine different application areas, leading to different threshold-based algorithms [30, 31, 63, 96, 3]. For example, the authors in [31] study top-k queries over multimedia repositories, whereas Marian *et al.* [96] study top-k query evaluation over web-accessible databases, including random accesses to score lists.

Following the same concept, related papers address the problem of top-k queries in peer-to-peer systems over vertically distributed data. In [23], Cao and Wang propose an algorithm called "Three-Phase Uniform Threshold" (TPUT) that aims to prune unnecessary data objects and it is guaranteed to terminate in three round-trips. Later, TPUT was improved by KLEE [97]. KLEE has two variants, one that requires three phases and another that only needs two round-trips. KLEE also provides mechanisms for trading performance with result quality, thus supporting approximate top-k retrieval.

For horizontally distributed data among peers, top-k query processing has been studied in only few works so far. Balke *et al.* [7] try to minimize the data object traffic induced by top-k processing. However, this approach requires that each query is processed by all super-peers, unless the exact same query reoccurs, which is unlikely as there is an infinite number of potential queries posed by different users. A similar approach for unstructured P2P systems is presented in [2], where the main technique is a variant of flooding, followed by a merging score-list step at intermediate peers. In [147], the authors rely on result caching to prune network paths and answer queries without contacting all peers. Their approach relies on caching techniques, therefore the performance is dependent on the query distribution. Even more important, they assume acyclic networks, which is restrictive for dynamic peer-to-peer networks. Hose *et al.* [72] construct routing filters in the form of histograms, in order to prune query paths and return approximate results. These filters are built on each peer progressively, as the peer communicates with other peers, using a query feedback approach. However this approach delivers approximate answers and the performance drops with increasing dimensionality since multi-dimensional histograms should be used. In [46], an approach is proposed that tries to minimize the users' waiting time of top-k results, at the expense of multiple phases of data transmission. Recently, Ryeng *et al.* [113] studied caching of top-k results and the use of remainder queries to answer future top-k queries.

In the area of peer-to-peer information retrieval, there exists some work that takes into account top-k queries. However this work is not entirely within the context of this chapter, as their main focus is on document retrieval and on defining an appropriate scoring function. For example, in [94, 95], Lu and Callan focus on search in a digital library context, using hierarchical P2P networks and propose result merging algorithms based on sampled documents from neighboring peers.

7.6 Summary

In this chapter, we study the challenging problem of efficient top-k query processing processing over multiple servers. The aim of distributed top-k query processing
is to forward the query only to the servers that store relevant data, in such a way
that the amount of transferred data is minimized. Towards this goal, we explore the
applicability of the skyline points for efficiently identifying the servers that store
relevant data. Then, a threshold-based algorithm is employed to return the result set
for any top-k query, while supporting a large class of scoring functions. Since the
number of contacted super-peer and the amount of transferred data are minimized,
distributed top-k query processing is performed efficiently even in the case where a
high degree of distribution is required.

Chapter 8
Summary

Abstract Query processing over multidimensional data in distributed systems is an important and emerging challenge. In this chapter, the most important findings in the book are summarized and future directions are outlined within the area of distributed query processing over multidimensional data.

8.1 Conclusions

In an increasingly high number of applications there is the need for storing large amounts of data in geographically distributed locations and involving a high number of computers. In such applications, a common requirement is to perform content-based search or advanced ranking through complex and costly query operators. In some setups, typically where one organization or company has full control, the challenge can be solved by using a number of data centers, each with a large number of computers. In this book, we have targeted application areas where this is not the case, and each participant enjoys a high degree of autonomy. Examples of such applications include scientific databases, bioinformatics applications, and retrieval systems storing multimedia contents. Quite often, these domains are distributed in their very nature; data is produced in a distributed way and cannot be assembled at one location, due to its massive volume and high rate of generation. Moreover, data is typically high-dimensional, which further complicates query processing.

The main challenge for the above problem is scalability with respect to the number of participating computers and the number of queries. The P2P paradigm respects the system's decentralized nature, avoids bottlenecks, and ensures fault-tolerance. More important, P2P systems are capable of handling very large number of nodes, as demonstrated by deployed systems in highly distributed settings. In order to ensure efficiency and effectiveness during query processing, a hybrid P2P architecture (super-peer) can be employed that harnesses the merits of both worlds: the advantages of P2P and the efficiency of centralized query processing.

In this book, a framework has been described for distributed query processing over multidimensional data using summary-based routing indices, and it has been illustrated how some advanced multidimensional query operators can be implemented within this framework. The use of summary-based routing indices facilitates the reduction of the number of super-peers that have to be involved in a particular query. Thus, the result is a highly scalable framework for processing queries involving a large number of physically distributed computers.

8.2 Future Directions

Even though several approaches have been proposed for efficient P2P query processing over multidimensional data, there are still challenging issues about P2P query processing that have not been studied sufficiently in the related literature.

A current trend in database research is efficient processing of probabilistic queries over uncertain data. In distributed environments, the uncertainty of the data can be caused by several different reasons. In some cases, the data can be uncertain, similar to the case of the centralized setting. Consider for example, noisy readings from sensors in sensor networks. On the other hand, the uncertainty in data is inherent in P2P systems due to the lack of central control that can verify the quality of the data. For example, in a P2P network, some peers may not always trust other peers, since some of them might act as cheaters. In this case, the querying peer may be associated with a value of trustworthiness that indicates the validity of data stored by different peers. P2P query processing over uncertain data is an important problem for many real-world distributed applications and efficient methods need to be developed.

Nowadays, there exist many different distributed systems that have characteristics that differ from P2P systems. Even though the framework presented in this book has been proposed in the context of P2P architectures, it can also be employed in other settings where high scalability and distribution are key issues. For example, several principles are applicable in other distributed systems where the data is distributed over autonomous servers such as grid systems, large-scale data centers or cloud computing infrastructures. In our future work, we plan to investigate the applicability of distributed query processing algorithms in other environments, such as cloud computing platforms.

Finally, in several distributed environments, such as mobile networks or cloud computing, users may be charged based on the amount of transferred data or the available bandwidth may be restricted. In a distributed environment, the performance of multidimensional query processing depends on the cardinality of the result set because at least these points have to be transferred to the querying peer. Moreover, the total number of transferred data points is much higher than the cardinality of the result set. An interesting direction of P2P query processing is to develop cost-efficient methods for computing an approximation of the query result set of high quality. The main goal is to process bandwidth-constrained queries by transferring only some of the local query result, but still the result set of the P2P query should be very similar to the exact result set.

References

1. R. Akbarinia, V. Martins, E. Pacitti, and P. Valduriez. Design and implementation of Atlas P2P architecture. In *Global Data Management*, 2006.
2. R. Akbarinia, E. Pacitti, and P. Valduriez. Reducing network traffic in unstructured P2P systems using top-k queries. *Distributed and Parallel Databases*, 19(2-3):67–86, 2006.
3. R. Akbarinia, E. Pacitti, and P. Valduriez. Best position algorithms for top-k queries. In *Proceedings of International Conference on Very Large Data Bases (VLDB)*, pages 495–506, 2007.
4. S. Androutsellis-Theotokis and D. Spinellis. A survey of peer-to-peer content distribution technologies. *ACM Computing Surveys*, 36(4):335–371, 2004.
5. W.-T. Balke, U. Güntzer, and J. X. Zheng. Efficient distributed skylining for web information systems. In *Proceedings of International Conference on Extending Database Technology (EDBT)*, pages 256–273, 2004.
6. W.-T. Balke and U. Gunzer. Multi-objective query processing for database systems. In *Proceedings of International Conference on Very Large Data Bases (VLDB)*, pages 936–947, 2004.
7. W.-T. Balke, W. Nejdl, W. Siberski, and U. Thaden. Progressive distributed top-k retrieval in peer-to-peer networks. In *Proceedings of IEEE International Conference on Data Engineering (ICDE)*, pages 174–185, 2005.
8. F. Banaei-Kashani and C. Shahabi. SWAM: A family of access methods for similarity-search in peer-to-peer data networks. In *Proceedings of International Conference on Information and Knowledge Management (CIKM)*, pages 304–313, 2004.
9. I. Bartolini, P. Ciaccia, V. Oria, and M. T. Ozsu. Flexible integration of multimedia sub-queries with qualitative preferences. *Multimedia Tools and Applications*, 33(3):275–300, 2007.
10. I. Bartolini, P. Ciaccia, and M. Patella. SaLSa: Computing the skyline without scanning the whole sky. In *Proceedings of International Conference on Information and Knowledge Management(CIKM)*, pages 405–414, 2006.
11. M. Batko, C. Gennaro, and P. Zezula. A scalable nearest neighbor search in P2P systems. In *International Workshop on Databases, Information Systems and Peer-to-Peer Computing (DBISP2P)*, pages 79–92, 2004.
12. M. Batko, D. Novak, F. Falchi, and P. Zezula. On scalability of the similarity search in the world of peers. In *Proceedings of International Conference on Scalable Information Systems (Infoscale)*, page 20, 2006.
13. M. Bawa, T. Condie, and P. Ganesan. LSH forest: Self-tuning indexes for similarity search. In *Proceedings of the International Conference on World Wide Web (WWW)*, pages 651–660, 2005.

14. J. L. Bentley, H. T. Kung, M. Schkolnick, and C. D. Thompson. On the average number of maxima in a set of vectors and applications. *Journal of the ACM (JACM)*, 25(4):536–543, 1978.

15. S. Berchtold, C. Böhm, D. A. Keim, and H.-P. Kriegel. A cost model for nearest neighbor search in high-dimensional data space. In *Proceedings of Symposium on Principles of Database Systems (PODS)*, pages 78–86, 1997.

16. K. S. Beyer, J. Goldstein, R. Ramakrishnan, and U. Shaft. When is "nearest neighbor" meaningful? In *Proceedings of the International Conference on Database Theory (ICDT)*, pages 217–235, 1999.

17. A. R. Bharambe, M. Agrawal, and S. Seshan. Mercury: supporting scalable multi-attribute range queries. In *Proceedings of ACM Conference on Applications, technologies, architectures, and protocols for computer communications (SIGCOMM)*, pages 353–366, 2004.

18. I. Bhattacharya, S. R. Kashyap, and S. Parthasarathy. Similarity searching in peer-to-peer databases. In *Proceedings of the International Conference on Distributed Computing Systems (ICDCS)*, pages 329–338, 2005.

19. E. Bingham and H. Mannila. Random projection in dimensionality reduction: Applications to image and text data. In *Proceedings of the International Conference on Knowledge discovery and data mining (SIGKDD)*, pages 245–250, 2001.

20. P. Boncz and C. Treijtel. AmbientDB: relational query processing in a P2P network. In *International Workshop on Databases, Information Systems and Peer-to-Peer Computing (DBISP2P)*, pages 153–168, 2003.

21. S. Borzsonyi, D. Kossmann, and K. Stocker. The skyline operator. In *Proceedings of IEEE International Conference on Data Engineering (ICDE)*, pages 421–430, 2001.

22. N. Bruno, S. Chaudhuri, and L. Gravano. Top-k selection queries over relational databases: Mapping strategies and performance evaluation. *ACM Transactions on Database Systems (TODS)*, 27(2):153–187, 2002.

23. P. Cao and Z. Wang. Efficient top-k query calculation in distributed networks. In *Proceedings of Annual ACM Symposium on Principles of Distributed Computing (PODC)*, pages 206–215, 2004.

24. M. J. Carey and D. Kossmann. On saying "Enough Already!" in SQL. In *Proceedings of International Conference on Management of Data (SIGMOD)*, pages 219–230, 1997.

25. K. Chakrabarti and S. Mehrotra. Local dimensionality reduction: A new approach to indexing high dimensional spaces. In *Proceedings of International Conference on Very Large Data Bases (VLDB)*, pages 89–100, 2000.

26. C. Y. Chan, H. V. Jagadish, K.-L. Tan, A. K. H. Tung, and Z. Zhang. Finding k-dominant skylines in high dimensional space. In *Proceedings of International Conference on Management of Data (SIGMOD)*, pages 503–514, 2006.

27. C. Y. Chan, H. V. Jagadish, K.-L. Tan, A. K. H. Tung, and Z. Zhang. On high dimensional skylines. In *Proceedings of International Conference on Extending Database Technology (EDBT)*, pages 478–495, 2006.

28. Y.-C. Chang, L. D. Bergman, V. Castelli, C.-S. Li, M.-L. Lo, and J. R. Smith. The onion technique: Indexing for linear optimization queries. In *Proceedings of International Conference on Management of Data (SIGMOD)*, pages 391–402, 2000.

29. S. Chaudhuri, N. N. Dalvi, and R. Kaushik. Robust cardinality and cost estimation for skyline operator. In *Proceedings of IEEE International Conference on Data Engineering (ICDE)*, page 64, 2006.

30. S. Chaudhuri and L. Gravano. Evaluating top-k selection queries. In *Proceedings of International Conference on Very Large Data Bases (VLDB)*, pages 397–410, 1999.

31. S. Chaudhuri, L. Gravano, and A. Marian. Optimizing top-k selection queries over multimedia repositories. *IEEE Transactions on Knowledge and Data Engineering (TKDE)*, 16(8):992–1009, 2004.

32. E. Chavez, G. Navarro, R. Baeza-Yates, and J. L. Marroquin. Searching in metric spaces. *ACM Computing Surveys*, 33(3):273–321, 2001.

33. C.-M. Chen and Y. Ling. A sampling-based estimator for top-k query. In *Proceedings of IEEE International Conference on Data Engineering (ICDE)*, pages 617–627, 2002.

34. L. Chen, B. Cui, and H. Lu. Constrained skyline query processing against distributed data sites. *IEEE Transactions on Knowledge and Data Engineering (TKDE)*, 23(2):204–217, 2011.

35. L. Chen, B. Cui, H. Lu, L. Xu, and Q. Xu. iSky: Efficient and progressive skyline computing in a structured P2P network. In *Proceedings of the International Conference on Distributed Computing Systems (ICDCS)*, pages 160–167, 2008.

36. J. Chomicki, P. Godfrey, J. Gryz, and D. Liang. Skyline with pre-sorting. In *Proceedings of IEEE International Conference on Data Engineering (ICDE)*, pages 717–719, 2003.

37. P. Ciaccia, M. Patella, and P. Zezula. M-tree: An efficient access method for similarity search in metric spaces. In *Proceedings of International Conference on Very Large Data Bases (VLDB)*, pages 426–435, 1997.

38. P. Ciaccia, M. Patella, and P. Zezula. A cost model for similarity queries in metric spaces. In *Proceedings of Symposium on Principles of Database Systems (PODS)*, pages 59–68, 1998.

39. I. Clarke, O. Sandberg, B. Wiley, and T. Hong. Freenet: A distributed anonymous information storage and retrieval system. In *Proceedings of the ICSI Workshop on Design Issues in Anonymity and Unobservability*, 2000.

40. A. Cosgaya-Lozano, A. Rau-Chaplin, and N. Zeh. Parallel computation of skyline queries. In *Proceedings of International Symposium on High Performance Computing Systems and Applications*, page 12, 2007.

41. A. Crainiceanu, P. Linga, J. Gehrke, and J. Shanmugasundaram. P-tree: A P2P index for resource discovery applications. In *Proceedings of the International Conference on World Wide Web (WWW)*, 2004.

42. A. Crespo and H. Garcia-Molina. Routing indices for peer-to-peer systems. In *Proceedings of the International Conference on Distributed Computing Systems (ICDCS)*, page 23, 2002.

43. B. Cui, L. Chen, L. Xu, H. Lu, G. Song, and Q. Xu. Efficient skyline computation in structured peer-to-peer systems. *IEEE Transactions on Knowledge and Data Engineering (TKDE)*, 21(7):1059–1072, 2009.

44. B. Cui, H. Lu, Q. Xu, L. Chen, Y. Dai, and Y. Zhou. Parallel distributed processing of constrained skyline queries by filtering. In *Proceedings of International Conference on Data Engineering (ICDE)*, pages 546–555, 2008.

45. A. Datta, M. Hauswirth, R. John, R. Schmidt, and K. Aberer. Range queries in trie-structured overlays. In *Proceedings of IEEE International Conference on Peer-to-Peer Computing*, pages 57–66, 2005.

46. W. K. Dedzoe, P. Lamarre, R. Akbarinia, and P. Valduriez. Asap top-k query processing in unstructured p2p systems. In *Peer-to-Peer Computing*, pages 1–10, 2010.

47. E. Dellis, A. Vlachou, I. Vladimirskiy, B. Seeger, and Y. Theodoridis. Constrained subspace skyline computation. In *Proceedings of International Conference on Information and Knowledge Management(CIKM)*, pages 415–424, 2006.

48. C. Doulkeridis, K. Nørvåg, and M. Vazirgiannis. DESENT: Decentralized and distributed semantic overlay generation in P2P networks. *IEEE Journal on Selected Areas in Communications (J-SAC)*, 25(1):25–34, 2007.

49. C. Doulkeridis, A. Vlachou, Y. Kotidis, and M. Vazirgiannis. Peer-to-peer similarity search in metric spaces. In *Proceedings of International Conference on Very Large Data Bases (VLDB)*, pages 986–997, 2007.

50. C. Doulkeridis, A. Vlachou, Y. Kotidis, and M. Vazirgiannis. Efficient range query processing in metric spaces over highly distributed data. *Distributed and Parallel Databases*, 26(2-3):155–180, 2009.

51. C. Doulkeridis, A. Vlachou, K. Nørvåg, Y. Kotidis, and M. Vazirgiannis. Multidimensional routing indices for efficient distributed query processing. In *Proceedings of International Conference on Information and Knowledge Management (CIKM)*, pages 1489–1492, 2009.

52. C. Doulkeridis, A. Vlachou, K. Nørvåg, Y. Kotidis, and M. Vazirgiannis. On the selectivity of multidimensional routing indices. In *Proceedings of International Conference on Information and Knowledge Management (CIKM)*, pages 109–118, 2010.

53. C. Doulkeridis, A. Vlachou, K. Nørvåg, and M. Vazirgiannis. Distributed semantic overlay networks. In *Handbook of Peer-to-Peer Networking*. Springer, 2011.

54. R. Fagin, A. Lotem, and M. Naor. Optimal aggregation algorithms for middleware. In *Proceedings of Symposium on Principles of Database Systems (PODS)*, pages 102–113, 2001.

55. F. Falchi, C. Gennaro, and P. Zezula. A content-addressable network for similarity search in metric spaces. In *International Workshop on Databases, Information Systems and Peer-to-Peer Computing (DBISP2P)*, pages 126–137, 2005.

56. C. Faloutsos and K.-I. Lin. Fastmap: A fast algorithm for indexing, data-mining and visualization of traditional and multimedia datasets. In *Proceedings of International Conference on Management of Data (SIGMOD)*, pages 163–174, 1995.

57. V. Gaede and O. Günther. Multidimensional access methods. *ACM Computing Surveys*, 30(2):170–231, 1998.

58. P. Ganesan, M. Bawa, and H. Garcia-Molina. Online balancing of range-partitioned data with applications to peer-to-peer systems. In *Proceedings of International Conference on Very Large Data Bases (VLDB)*, pages 444–455, 2004.

59. Y. Gao, G. Chen, L. Chen, and C. Chen. Parallelizing progressive computation for skyline queries in multi-disk environment. In *Proceedings of International Conference on Database and Expert Systems Applications (DEXA)*, pages 697–706, 2006.

60. C. Gkantsidis, M. Mihail, and A. Saberi. Random walks in peer-to-peer networks. In *Proceedings of Annual Joint Conference of Computer and Communications Societies (INFO-COM)*, 2004.

61. P. Godfrey. Skyline cardinality for relational processing. In *Proceedings of Foundations of Information and Knowledge Systems (FoIKS)*, pages 78–97, 2004.

62. P. Godfrey, R. Shipley, and J. Gryz. Maximal vector computation in large data sets. In *Proceedings of International Conference on Very Large Data Bases (VLDB)*, pages 229–240, 2005.

63. U. Güntzer, W.-T. Balke, and W. Kießling. Optimizing multi-feature queries for image databases. In *Proceedings of International Conference on Very Large Data Bases (VLDB)*, pages 419–428, 2000.

64. A. Gupta, D. Agrawal, and A. E. Abbadi. Approximate range selection queries in peer-to-peer systems. In *Proceedings of Biennial Conference on Innovative Data Systems Research (CIDR)*, 2003.

65. A. Guttman. R-trees: A dynamic index structure for spatial searching. In *Proceedings of International Conference on Management of Data (SIGMOD)*, pages 47–57, 1984.

66. A. Y. Halevy, Z. G. Ives, J. Madhavan, P. Mork, D. Suciu, and I. Tatarinov. The Piazza peer data management system. *IEEE Transactions on Knowledge and Data Engineering (TKDE)*, 16(7):787–798, 2004.

67. J. O. Hauglid, N. H. Ryeng, and K. Nørvåg. The DASCOSA-DB grid database system. In *Grid and Cloud Database Management*, pages 87–106. Springer, 2011.

68. A. Hinneburg, C. C. Aggarwal, and D. A. Keim. What is the nearest neighbor in high dimensional spaces? In *Proceedings of International Conference on Very Large Data Bases (VLDB)*, pages 506–515, 2000.

69. G. R. Hjaltason and H. Samet. Ranking in spatial databases. In *Proceedings of the 4th International Symposium on Advances in Spatial Databases (SSD)*, pages 83–95, 1995.

70. G. R. Hjaltason and H. Samet. Distance browsing in spatial databases. *ACM Transactions on Database Systems (TODS)*, 24(2):265–318, 1999.

71. G. R. Hjaltason and H. Samet. Index-driven similarity search in metric spaces. *ACM Transactions on Database Systems (TODS)*, 28(4):517–580, 2003.

72. K. Hose, M. Karnstedt, K.-U. Sattler, and D. Zinn. Processing top-N queries in P2P-based web integration systems with probabilistic guarantees. In *Proceedings of International Workshop on Web and Databases (WebDB)*, pages 109–114, 2005.

73. K. Hose, C. Lemke, and K.-U. Sattler. Processing relaxed skylines in PDMS using distributed data summaries. In *Proceedings of International Conference on Information and Knowledge Management(CIKM)*, pages 425–434, 2006.

74. K. Hose, C. Lemke, and K.-U. Sattler. Maintenance strategies for routing indexes. *Distributed and Parallel Databases*, 26(2-3):231–259, 2009.

75. V. Hristidis, N. Koudas, and Y. Papakonstantinou. PREFER: A system for the efficient execution of multi-parametric ranked queries. In *Proceedings of International Conference on Management of Data (SIGMOD)*, pages 259–270, 2001.

76. Z. Huang, C. Jensen, H. Lu, and B.-C. Ooi. Skyline queries against mobile lightweight devices in MANETs. In *Proceedings of IEEE International Conference on Data Engineering (ICDE)*, page 66, 2006.

77. R. Huebsch, J. M. Hellerstein, N. Lanham, B. T. Loo, S. Shenker, and I. Stoica. Querying the internet with PIER. In *Proceedings of International Conference on Very Large Data Bases (VLDB)*, pages 321–332, 2003.

78. I. F. Ilyas, G. Beskales, and M. A. Soliman. A survey of top-k query processing techniques in relational database systems. *ACM Computing Surveys*, 40(4):1–58, 2008.

79. G. Y. J. Lee and S. Hwang. Telescope: Zooming to interesting skylines. In *Proceedings of International Conference on Database Systems for Advanced Applications (DASFAA)*, 2007.

80. H. V. Jagadish, B. C. Ooi, K.-L. Tan, C. Yu, and R. Zhang. iDistance: An adaptive B^+-tree based indexing method for nearest neighbor search. *ACM Transactions on Database Systems (TODS)*, 30(2):364–397, June 2005.

81. H. V. Jagadish, B. C. Ooi, and Q. H. Vu. Baton: A balanced tree structure for peer-to-peer networks. In *Proceedings of International Conference on Very Large Data Bases (VLDB)*, pages 661–672, 2005.

82. H. V. Jagadish, B. C. Ooi, Q. H. Vu, R. Zhang, and A. Zhou. VBI-tree: A peer-to-peer framework for supporting multi-dimensional indexing schemes. In *Proceedings of IEEE International Conference on Data Engineering (ICDE)*, page 34, 2006.

83. W. Jin, A. K. H. Tung, M. Ester, and J. Han. On efficient processing of subspace skyline queries on high dimensional data. In *Proceedings of the 19th International Conference on Scientific and Statistical Database Management (SSDBM)*, page 12, 2007.

84. P. Kalnis, W. S. Ng, B. C. Ooi, and K.-L. Tan. Answering similarity queries in peer-to-peer networks. *Information Systems Journal*, 31(1):57–72, 2006.

85. V. Koltun and C. H. Papadimitriou. Approximately dominating representatives. *Theoretical Computer Science*, 371(3):148–154, 2007.

86. D. Kossmann, F. Ramsak, and S. Rost. Shooting stars in the sky: an online algorithm for skyline queries. In *Proceedings of International Conference on Very Large Data Bases (VLDB)*, pages 275–286, 2002.

87. W.-S. Ku, R. Zimmermann, C.-N. Wan, and H. Wang. Maple: A mobile scalable P2P nearest neighbor query system for location-based services. In *Proceedings of IEEE International Conference on Data Engineering (ICDE)*, page 160, 2006.

88. J. Kubiatowicz, D. Bindel, Y. Chen, S. E. Czerwinski, P. R. Eaton, D. Geels, R. Gummadi, S. C. Rhea, H. Weatherspoon, W. Weimer, C. Wells, and B. Y. Zhao. Oceanstore: An architecture for global-scale persistent storage. In *Proceedings of International Conference on Architectural Support for Programming Languages and Operating Systems (ASPLOS)*, pages 190–201, 2000.

89. H. T. Kung, F. Luccio, and F. P. Preparata. On finding the maxima of a set of vectors. *Journal of the ACM (JACM)*, 22(4):469–476, 1975.

90. K. Lee, B. Zheng, H. Li, and W.-C. Lee. Approaching the skyline in Z order. In *Proceedings of International Conference on Very Large Data Bases (VLDB)*, pages 279–290, 2007.

91. X. Lin, Y. Yuan, Q. Zhang, and Y. Zhang. Selecting stars: The k most representative skyline operator. In *Proceedings of IEEE International Conference on Data Engineering (ICDE)*, 2007.

92. B. Liu, W.-C. Lee, and D. L. Lee. Supporting complex multi-dimensional queries in P2P systems. In *Proceedings of the International Conference on Distributed Computing Systems (ICDCS)*, pages 155–164, 2005.

93. E. Lo, K. Y. Yip, K.-I. Lin, and D. W. Cheung. Progressive skylining over web-accessible databases. *Data and Knowledge Engineering (DKE)*, 57(2):122–147, 2006.

94. J. Lu and J. Callan. Merging retrieval results in hierarchical peer-to-peer networks. In *Proceedings of the ACM International Conference on Research and Development in Information Retrieval (SIGIR)*, pages 472–473, 2004.

95. J. Lu and J. Callan. Federated search of text-based digital libraries in hierarchical peer-to-peer networks. In *Proceedings of European Conference on IR Research (ECIR)*, pages 52–66, 2005.

96. A. Marian, N. Bruno, and L. Gravano. Evaluating top-k queries over web-accessible databases. *ACM Transactions on Database Systems (TODS)*, 29(2):319–362, 2004.

97. S. Michel, P. Triantafillou, and G. Weikum. KLEE: A framework for distributed top-k query algorithms. In *Proceedings of International Conference on Very Large Data Bases (VLDB)*, pages 637–648, 2005.

98. M. Morse, J. M. Patel, and H. V. Jagadish. Efficient skyline computation over low-cardinality domains. In *Proceedings of International Conference on Very Large Data Bases (VLDB)*, pages 267–278, 2007.

99. W. S. Ng, B. C. Ooi, K.-L. Tan, and A. Zhou. PeerDB: A P2P-based system for distributed data sharing. In *Proceedings of International Conference on Data Engineering (ICDE)*, 2003.

100. D. Novak and P. Zezula. M-Chord: A scalable distributed similarity search structure. In *Proceedings of International Conference on Scalable Information Systems (Infoscale)*, page 19, 2006.

101. N. Ntarmos, T. Pitoura, and P. Triantafillou. Range query optimization leveraging peer heterogeneity. In *International Workshop on Databases, Information Systems and Peer-to-Peer Computing (DBISP2P)*, 2005.

102. D. Papadias, Y. Tao, G. Fu, and B. Seeger. An optimal and progressive algorithm for skyline queries. In *Proceedings of International Conference on Management of Data (SIGMOD)*, pages 443–454, 2003.

103. D. Papadias, Y. Tao, G. Fu, and B. Seeger. Progressive skyline computation in database systems. *ACM Transactions on Database Systems*, 30(1):41–82, 2005.

104. J. Pei, A. W.-C. Fu, X. Lin, and H. Wang. Computing compressed multidimensional skyline cubes efficiently. In *Proceedings of IEEE International Conference on Data Engineering (ICDE)*, 2007.

105. J. Pei, W. Jin, M. Ester, and Y. Tao. Catching the best views of skyline: A semantic approach based on decisive subspaces. In *Proceedings of International Conference on Very Large Data Bases (VLDB)*, pages 253–264, 2005.

106. S. Ratnasamy, P. Francis, M. Handley, R. Karp, and S. Schenker. A scalable content-addressable network. In *Proceedings of ACM Conference on Applications, technologies, architectures, and protocols for computer communications (SIGCOMM)*, pages 161–172, 2001.

107. J. Risson and T. Moors. Survey of research towards robust peer-to-peer networks: Search methods. *Computer Networks*, 50(17):3485–3521, 2006.

108. J. B. Rocha-Junior, A. Vlachou, C. Doulkeridis, and K. Nørvåg. AGiDS: A grid-based strategy for distributed skyline query processing. In *Proceedings of International Conference on Data Management in Grid and Peer-to-Peer Systems (Globe)*, pages 12–23, 2009.

109. J. B. Rocha-Junior, A. Vlachou, C. Doulkeridis, and K. Nørvåg. Efficient execution plans for distributed skyline query processing. In *Proceedings of International Conference on Extending Database Technology (EDBT)*, pages 271–282, 2011.

110. P. Rodríguez-Gianolli, M. Garzetti, L. Jiang, A. Kementsietsidis, I. Kiringa, M. Masud, R. J. Miller, and J. Mylopoulos. Data sharing in the Hyperion peer database system. In *Proceedings of International Conference on Very Large Data Bases (VLDB)*, 2005.

111. N. Roussopoulos, S. Kelley, and F. Vincent. Nearest neighbor queries. In *Proceedings of International Conference on Management of Data (SIGMOD)*, pages 71–79, 1995.

112. A. Rowstron and P. Druschel. Pastry: Scalable, distributed object location and routing for large-scale peer-to-peer systems. In *Proceedings of the International Middleware Conference*, pages 329–350, 2001.

113. N. H. Ryeng, A. Vlachou, C. Doulkeridis, and K. Nørvåg. Efficient distributed top-k query processing with caching. In *Interantional Conference on Database Systems for Advanced Applications (DASFAA)*, pages 280–295, 2011.

114. O. D. Sahin, F. Emekçi, D. Agrawal, and A. E. Abbadi. Content-based similarity search over peer-to-peer systems. In *International Workshop on Databases, Information Systems and Peer-to-Peer Computing (DBISP2P)*, pages 61–78, 2004.

115. O. D. Sahin, A. Gupta, D. Agrawal, and A. E. Abbadi. A peer-to-peer framework for caching range queries. In *Proceedings of IEEE International Conference on Data Engineering (ICDE)*, page 165, 2004.

116. H. Samet. *Foundations of Multidimensional and Metric Data Structures (The Morgan Kaufmann Series in Computer Graphics and Geometric Modeling)*. Morgan Kaufmann Publishers Inc., 2005.

117. S. Saroiu, P. K. Gummadi, and S. D. Gribble. A measurement study of peer-to-peer file sharing systems. In *Proceedings of the Multimedia Computing and Networking*, 2002.

118. T. Seidl and H.-P. Kriegel. Efficient user-adaptable similarity search in large multimedia databases. In *Proceedings of International Conference on Very Large Data Bases (VLDB)*, pages 506–515, 1997.

119. T. Seidl and H.-P. Kriegel. Optimal multi-step k-nearest neighbor search. In *Proceedings of International Conference on Management of Data (SIGMOD)*, pages 154–165, 1998.

120. Y. Shu, B. C. Ooi, K.-L. Tan, and A. Zhou. Supporting multi-dimensional range queries in peer-to-peer systems. In *Proceedings of IEEE International Conference on Peer-to-Peer Computing*, pages 173–180, 2005.

121. I. Stoica, R. Morris, D. Karger, M. F. Kaashoek, and H. Balakrishnan. Chord: A scalable peer-to-peer lookup service for internet applications. In *Proceedings of ACM Conference on Applications, technologies, architectures, and protocols for computer communications (SIGCOMM)*, pages 149–160, 2001.

122. K.-L. Tan, P.-K. Eng, and B. C. Ooi. Efficient progressive skyline computation. In *Proceedings of International Conference on Very Large Data Bases (VLDB)*, pages 301–310, 2001.

123. C. Tang, Z. Xu, and S. Dwarkadas. Peer-to-peer information retrieval using self-organizing semantic overlay networks. In *Proceedings of ACM Conference on Applications, technologies, architectures, and protocols for computer communications (SIGCOMM)*, pages 175–186, 2003.

124. Y. Tao, X. Xiao, and J. Pei. SUBSKY: Efficient computation of skylines in subspaces. In *Proceedings of IEEE International Conference on Data Engineering (ICDE)*, page 65, 2006.

125. Y. Tao, X. Xiao, and J. Pei. Efficient skyline and top-k retrieval in subspaces. *IEEE Transactions on Knowledge and Data Engineering (TKDE)*, 19(8):1072–1088, 2007.

126. N. E. Taylor and Z. G. Ives. Reliable storage and querying for collaborative data sharing systems. In *Proceedings of International Conference on Data Engineering (ICDE)*, 2010.

127. Y. Theodoridis and T. K. Sellis. A model for the prediction of R-tree performance. In *Proceedings of Symposium on Principles of Database Systems (PODS)*, pages 161–171, 1996.

128. A. Vlachou, C. Doulkeridis, Y. Kotidis, and K. Nørvåg. Reverse top-k queries. In *Proceedings of IEEE International Conference on Data Engineering (ICDE)*, pages 365–376, 2010.

129. A. Vlachou, C. Doulkeridis, Y. Kotidis, and K. Nørvåg. Monochromatic and bichromatic reverse top-k queries. *IEEE Transactions on Knowledge and Data Engineering (TKDE)*, 23(8):1215–1229, 2011.

130. A. Vlachou, C. Doulkeridis, Y. Kotidis, and M. Vazirgiannis. SKYPEER: Efficient subspace skyline computation over distributed data. In *Proceedings of IEEE International Conference on Data Engineering (ICDE)*, pages 416–425, 2007.

131. A. Vlachou, C. Doulkeridis, and K. Nørvåg. Monitoring reverse top-k queries over mobile devices. In *Proceedings of the International Workshop on Data Engineering for Wireless and Mobile Access (MobiDE)*, 2011.

132. A. Vlachou, C. Doulkeridis, K. Nørvåg, and Y. Kotidis. Identifying the most influential data objects with reverse top-k queries. *PVLDB*, 3(1):364–372, 2010.

133. A. Vlachou, C. Doulkeridis, K. Nørvåg, and M. Vazirgiannis. On efficient top-k query processing in highly distributed environments. In *Proceedings of International Conference on Management of Data (SIGMOD)*, pages 753–764, New York, NY, USA, 2008.

134. A. Vlachou and K. Nørvåg. Bandwidth-constrained distributed skyline computation. In *Proceedings of the International Workshop on Data Engineering for Wireless and Mobile Access (MobiDE)*, pages 17–24, 2009.

135. A. Vlachou and M. Vazirgiannis. Ranking the sky: Discovering the importance of skyline points through subspace dominance relationships. *Data and Knowledge Engineering (DKE)*, 69(9):943–964, 2010.

136. S. Wang, B. C. Ooi, A. K. H. Tung, and L. Xu. Efficient skyline query processing on peer-to-peer networks. In *Proceedings of IEEE International Conference on Data Engineering (ICDE)*, pages 1126–1135, 2007.

137. S. Wang, Q. H. Vu, B. C. Ooi, A. K. Tung, and L. Xu. Skyframe: a framework for skyline query processing in peer-to-peer systems. *VLDB Journal*, 18(1):345–362, 2009.

138. P. Wu, C. Zhang, Y. Feng, B. Y. Zhao, D. Agrawal, and A. E. Abbadi. Parallelizing skyline queries for scalable distribution. In *Proceedings of International Conference on Extending Database Technology (EDBT)*, pages 112–130, 2006.

139. T. Xia and D. Zhang. Refreshing the sky: The compressed skycube with efficient support for frequent updates. In *Proceedings of International Conference on Management of Data (SIGMOD)*, pages 491–502, 2006.

140. T. Xia, D. Zhang, and Y. Tao. On skylining with flexible dominance relation. In *Proceedings of IEEE International Conference on Data Engineering (ICDE)*, 2008.

141. B. Yang and H. Garcia-Molina. Designing a super-peer network. In *Proceedings of IEEE International Conference on Data Engineering (ICDE)*, pages 49–, 2003.

142. K. Yi, H. Yu, J. Yang, G. Xia, and Y. Chen. Efficient maintenance of materialized top-k views. In *Proceedings of IEEE International Conference on Data Engineering (ICDE)*, pages 189–200, 2003.

143. C. Yu, S. Bressan, B. C. Ooi, and K.-L. Tan. Querying high-dimensional data in single-dimensional space. *VLDB Journal*, 13(2):105–119, 2004.

144. C. Yu, B. C. Ooi, K.-L. Tan, and H. V. Jagadish. Indexing the distance: An efficient method to KNN processing. In *Proceedings of International Conference on Very Large Data Bases (VLDB)*, 2001.

145. Y. Yuan, X. Lin, Q. Liu, W. Wang, J. Yu, and Q. Zhang. Efficient computation of the skyline cube. In *Proceedings of International Conference on Very Large Data Bases (VLDB)*, pages 241–252, 2005.

146. B. Zhao, J. Kubiatowicz, and A. Joseph. Tapestry: An infrastructure for fault-tolerant wide-area location and routing. Technical report, U.C.Berkeley, 2001.

147. K. Zhao, Y. Tao, and S. Zhou. Efficient top-k processing in large-scaled distributed environments. *Data and Knowledge Engineering (DKE)*, 63(2):315–335, 2007.

148. L. Zhu, Y. Tao, and S. Zhou. Distributed skyline retrieval with low bandwidth consumption. *IEEE Transactions on Knowledge and Data Engineering (TKDE)*, 21(3):384–400, 2009.